Understanding
Modern Health Care

Understanding
Modern Health Care

The Wonders We Created and the Potholes We Dug

STEVE FREDMAN, M.D.

ISBN: 978-0-578-88317-5
Ebook ISBN: 978-0-578-91307-0

Manufactured in the United States of America.

Produced by Dean Burrell
Design by Maureen Forys, Happenstance Type-O-Rama

Front cover illustration © Shutterstock/Vladimir Ischuk
Back cover photograph © Shutterstock/vchal

10 9 8 7 6 5 4 3 2 1

To Marion, Peter, Arden, Juliana, and Gabe with love.
Thanks for your support. I learned from all of you.

ACKNOWLEDGMENTS

I want to thank the people who read what I wrote and made me understand what I had to do to make them understand. In alphabetical order (by last name): Burt Calder, Essia Cartoon-Fredman, Susanna Cohen, Felix Davis, Angela Engel, Jeremy Evnine, Rachel Evnine, Anabel Fredman, Celia Fredman, Gabe Fredman, Howard Fredman, Juliana Fredman, Marion Fredman, Byron Hann, Eleanor Parks, Cinda Pearlman, Alan Rinzler, Joy Sterneck. A special thanks to RoseAnn DeMoro.

CONTENTS

INTRODUCTION

A little over a century ago, my six-year-old dad and his family lived in a small, wooden, dirt-floored cottage in a shtetl that straddled one of the main Ukrainian-Russian east-west highways. In 1914, the First World War started. The Russian army attacked Germany and fell into a trap. The Russian Second Army was virtually destroyed at the Battle of Tannenberg, and thousands of the surviving soldiers retreated. When they came through my father's town, the fleeing Cossacks burned the family home to the ground. During the subsequent war years, the family crowded into one of the remaining cabins on the edge of the village. It was owned by an elderly Ukrainian who hadn't left for mother Russia with his family.

During the war, no one bathed or boiled their clothes. Everyone's garments and bedding contained body lice. One winter there was a typhus outbreak. The infectious disease is caused by a tiny bacterium (rickettsia) that lives in the lice. When the creatures defecate, their droppings itch. People scratch, tear their skin, and bacteria enter their bodies. One to two weeks later, the aching starts. Many become quite ill. They have chills, high fevers, an unremitting headache, and exhaustion. When my grandmother became feverish, she was also confused. A Russian army nurse who was making the rounds came by. The family was unable to hide the sick woman and the nurse summoned a wagon. It took my grandmother to the schoolhouse, the large hall full of beds where most died. My grandfather watched and cried as they carted her away.

"During the eight years between 1917 and 1925, more than 25 million people living in Russia developed epidemic typhus, and three million died." Some claim epidemic typhus has caused more deaths than all the wars in history. My father always remembered his boyhood, and when I chose to go to medical school, he shrugged. Based on what he witnessed, he believed doctors know how to recognize and diagnose illness, but that's all they can do. (In the 21st century, typhus is easily cured and prevented with the antibiotic doxycycline.)

The human body knows how to mend itself and fight off infections, and there have always been healers and helpers. The first sign of civilization, according to anthropologist Margaret Mead, was a femur (thighbone) that had been fractured and healed. Repair and restoration takes time. Without help creatures with broken legs can't escape danger and don't survive.

Prior to the 1900s, mankind didn't have the ability and tools needed to cure the lame and blind, turn around a lethal infection, remove a cancer, or give someone a new heart or kidney. The needed drugs, devices, and skills—"health care"—were created (or transformed) during the last 120 years. It's a gift we received because we were born in the 20th and 21st centuries. It may be as common as the iPhone, the airplane, or the Internet. We may take it for granted and feel like it has always been and always will be available when we need it.

But it has become expensive. In the last fifty years, some of us have had to deal with costly insurance, obscenely priced medications, and outrageous hospital bills. Many of the people in charge don't believe health care is or should be a shared responsibility.

The authors of the Declaration of Independence didn't think health care was an "unalienable right that was endowed by our creator" and health care wasn't one of the many rights that were added to the nation's Constitution in 1791.

Back then, nursing care supported the ill and sped their recovery. Amputations prevented some deaths. But most of the treatments doctors employed were pretty awful. Consider—the December morning in 1799 when sixty-seven-year-old George Washington awoke desperately ill. He was retired and lived at Mt. Vernon. The previous day, Washington felt well and went out in the snow to "mark trees that were to be cut down." Upon awakening on the day in question, he couldn't talk and had trouble breathing. His wife, Martha, sent for one doctor, then another. She and her husband were two of the country's richest people and obviously didn't need subsidized care.

During the day, three prominent physicians came to their home and plied their trade. The doctors were among the country's best and they worked hard. On four occasions, they bled the sick man and removed a lot of blood. His throat was swabbed, he gargled, his feet were covered with wheat bran, and he was given an emetic to induce vomiting. Nothing worked. When Washington's breathing got worse, he dressed, thanked his three doctors, and made arrangements for his burial. That night, he died. (As related by his secretary Tobias Lear)

Before 1800, the educated elites relied on the teachings of the ancients, like the Greek physician Hippocrates, who believed that illness was "due to an imbalance of blood, phlegm, black bile, and yellow bile," and the Roman Galen who dissected monkeys and wrote about their anatomy.

Mankind was not aware of the microscopic creatures who lived in, around, and on us until the late 1700s.

During the 1800s we gradually learned about their existence. We started believing and understanding that they were the source of many of our maladies, and we began to take precautions.

In the 1900s our abilities exploded: We learned how to safely transfuse blood. Hormones were isolated. Antibiotics and drugs that fought viruses and parasites were developed. Experts learned and taught others how to replace eye lenses that were opaque. Vaccines were crafted. Thousands of medical gadgets were devised. Surgeons were taught how to proceed after they cut a person open, and a large number of effective drugs became available.

In 1965, more than 100 million Americans were introduced to socialized medicine—Medicare and Medicaid. Most loved it.

In 2003, the entire human genome was "sequenced." Scientists determined the exact order, the way the 3 billion pairs of human DNA nucleotides (building blocks) lined up, and our ability to attack and "cure" genetic conditions got a big boost. The push and pull between medical care as a shared endeavor or a wealth-producing commodity started in the 1900s. It intensified over time. In the last half of the 20th century, "health care" increasingly became a major part of the U.S. economy and obstacles and inequalities were created. This book seeks to make sense of the wonders that were developed and the challenges we face.

In the pages that follow, I'd like you to accompany me up the miraculous, tortuous road medicine has traveled during the last two hundred years, and get a more detailed understanding of what I'm talking about.

CHAPTER ONE

Awakening to the Microscopic World

M odern health care's creation was triggered by the observations of a Dutch man named van Leeuwenhoek. Like the fictional Gulliver, he became the first to make the voyage, the first to gaze at the unknown world through a powerful lens.

A contemporary of Rembrandt and Vermeer, Leeuwenhoek was born in 17th-century Delft. It's a town in western Holland known for cool, foggy summer mornings, numerous boat-filled canals, wide streets connected by wooden bridges, and blue and white pottery. In his day, horses and carts clattered across the stones in front of a large open-air market, narrow rows of houses surrounded the town square, and food and wood were weighed before they were sold. Leeuwenhoek's mother came from a well-to-do brewer's family and van Leeuwenhoek first worked as a draper's apprentice. While there, he used the lenses of the day to check the quality of a fabric's thread. Later in life, he was politically active. He became a civil servant and was a chamberlain of one of the assembly chambers at city hall. At age forty, he made one, and later many, incredibly powerful, tiny magnifying lenses. Once he had created the devices, he started exploring the microscopic world, and he saw sights that had never before been seen or suspected. He drew pictures and sent them to the National Geographic of his day, the Royal Society. His images of bacteria, red blood cells, and sperm seemed fictional to some contemporaries who looked through ordinary polished and ground glass lenses. Others believed. During his life, Leeuwenhoek made an additional five hundred magnifiers. One person, then another, became aware of a microscopic world and learned it was often unfriendly. To this day no one really knows how Leeuwenhoek made his lenses. His process died with him.

Back then, 30 percent of those who got smallpox died, and it was known that survivors won't get it again. That's why some in ancient China and Africa blew crusts of a diseased person's scab up the nose of an uninfected person. They hoped the illness they were causing would be mild, or at worst, it wouldn't be deadly.

In 1796, Edward Jenner, a British doc, proved there was a safer way to prevent the disease. He heard that milkmaids who were infected by cowpox didn't develop smallpox, and he checked it out. He took material from the pussy scabs on a young woman's hand and "inserted it into small incisions he made in a boy's skin." Much as cats and tigers are members of the same species, the viruses that cause cowpox and smallpox are related. Each can cause pustular lesions. People who develop cowpox sometimes run a fever and are sick for a week, but the illness is mild, and when a person recovers (or is vaccinated) their body is protected from the oft lethal disease—smallpox.

Jenner submitted his findings to the Royal Society and they were rejected, so he self-published and became famous. Thomas Jefferson and James Madison read about his findings, and in 1813 Congress passed the Vaccine Act.

In 1853, the British Parliament made childhood vaccination with modified cowpox compulsory.

After widespread immunization contained the illness, people in the U.S. stopped vaccinating. In the 19th century, there were outbreaks, and states attempted to enforce existing laws or pass new ones. The disease finally disappeared from North America in 1952 and from Europe in 1953. As recently as 1967 (according to the CDC), 10 to 15 million people in Africa, Asia, Indonesia, and Brazil were contracting smallpox each year. That year, 2 million died and many were scarred for life. The World Health Organization started a program of worldwide vaccinations.

Their efforts to eliminate the terrible disease seem to have succeeded. The bug's last known "natural" victim was infected in 1977.

In 1848, hand washing was little more than a cultural or religious ritual. No one (best I can tell) connected "germs" and sanitation with infectious illnesses. That year, a Hungarian physician, Ignaz Semmelweis, was working at a hospital in Vienna and was troubled. Women whose babies were delivered by doctors and medical students developed a fever and died four times more often than women whose babies who were delivered by midwives.

Semmelweis investigated and learned that the medical students in question came from the dissecting room to the maternity ward without cleaning their hands. He introduced hand washing and the death rate plummeted. Unfortunately, his fellow physicians continued to believe that the high rate of childbed fever was due to "miasmas," clouds of invisible matter, and Semmelweis lost his job. The son of a prosperous grocer, he returned to Budapest, his hometown. In 1881, he published a book on "childbed fever." When he was in his late forties he was overcome by paranoia and dementia and he was committed to a psychiatric institution. It took a generation before his teachings were widely accepted.

While Semmelweis was investigating sanitation in Vienna, Louis Pasteur, a French chemist, was graduating and becoming a researcher. When he was young, Pasteur was an average student who loved to draw and paint. Then he got his act together and "won first prize in physics." He eventually studied chemistry and physics at the prestigious Ecole Normale.

At age twenty-six, Louis married twenty-three-year-old Marie Laurent. "According to legend he spent the morning of his wedding day in the lab and became so wrapped up in what he was doing that he had to be reminded to go to church."

Pasteur was thirty-two when he became a professor of chemistry at the university in Lille, a market city near the Belgian border whose streets were paved with stones and whose skies were often gray and rainy. In Lille, and three years later in Paris, Pasteur showed his fellow scientists that living organisms, bacteria, caused fermentation. We call it the germ theory. In 1863, working for the French emperor Napoleon III, Pasteur proved it was possible to eradicate harmful bacteria at a temperature well below boiling. He prevented wine from contamination by heating fermented grape juice to 50–60°C (120–140°F). We call the process pasteurization.

In 1879, he and his assistants injected chicken cholera bacteria into some of his birds. The germs had been sitting on the shelf for a while, the infections they caused were mild, and the infected chickens were subsequently resistant to the bug. Pasteur realized it was possible to weaken a pathogen to the point where it wasn't harmful but still triggered an immune response. He exploited the phenomenon to develop vaccines for chicken cholera and anthrax.

In 1885, a rabid dog bit a nine-year-old French child. We now know that after it enters a person's body, the virus that causes rabies infects an axon, the "long slender projections of nerve cells that conduct electrical impulses." The infectious agent then travels up the axon to the brain and eventually kills the person or animal.

The oft-repeated story says the young man was bitten fifteen times, and two days later his mother came knocking. Pasteur had for some time been injecting the agent that caused rabies into rabbits and had created pieces of spinal cord that were infectious. He had proven in dogs that when he dried the infected tissue in air it gradually became less virulent. Pasteur injected the boy with a series of fourteen increasingly virulent fragments of dried homogenized rabbit spinal cord. The boy survived and Pasteur's fame grew. Doctors started using similar extracts to treat people who were bitten by a rabid creature. The vaccines of the 21st century contain inactivated virus that was grown in human or chick embryo cells.

The rabies virus is still responsible for the deaths of 59,000 humans a year. Ninety percent of the cases in Africa and Asia are caused by dogs. In this country, we worry about bats and wild animals, and the U.S. has fewer than five confirmed cases a year. In his later years Pasteur had a series of strokes and he died when he was in his 70s.

In the early 1800s, the quality of microscopes was variable. Then a few craftsmen started making clear, powerful magnifying lenses. One of them, Carl Zeiss, came from a German family of artisans and he apprenticed with a maker of fine tools. In 1846, he opened a workshop in Jena, a river valley town in the "green heart" region of eastern Germany. The first dozen years, his technicians, under the supervision of a short-tempered, authoritarian foreman, made single lens precision microscopes. Eleven years later, Zeiss introduced scopes with two lenses. Scientists could now look into the upper curved glass, peer down a tube, and view an object that was just below a second lens. With the help of Ernst Abbe, a mathematician, the company used calculations to determine the optical characteristics of their lenses, and it improved the illumination system. Smaller and smaller objects came into view. Doctors from Germany and beyond bought one of their scopes. Carl's first wife died shortly after she gave birth to their first son. She was twenty-two at the time. Carl married two more times and outlived one of the women. All three of them were, in his words, "spiritually very much country folk."

Robert Koch, the German "father" of the germ theory, once wrote that his Zeiss scope was responsible for a large part of his success. Koch was born twenty-one years after Pasteur. He was a gifted child and could read a newspaper when he was five. In Germany he ran a medical practice and spent hours peering into a microscope. When he was a district medical officer, he investigated a pasture where the cows that ate the grass got sick and died. He collected blood from one of the dead animals, injected it into a mouse, and the rodent died. Koch found rod-shaped microscopic creatures in the soil, grew them in a rabbit's eye, and allowed them to dry out. They looked innocuous, but they were just dormant. When their survival was threatened, the bacteria surrounded themselves with a protein coat, became spores, and vegetated. They were able to endure harsh conditions, and when conditions permitted, they emerged. In the 20th century, these spores—anthrax—became one of the agents bioterrorists use.

Koch's life as a researcher started after he returned from the 1870 war with France. When the conflict started, Koch, "a five-and-a-half-foot tall man with a stern face and thin high voice," tried to become an army physician. He was rejected because he was nearsighted. As the conflict wore on, he reapplied and became a military doctor. He was with the German troops that besieged Orleans. It's the city on the Loire River where, in 1429, Joan of Arc famously fought the English. Koch was troubled by the damaged bodies he had to deal with. He once observed that in wartime human life becomes "worthless."

Years later, Koch was a famed researcher. When he was forty-seven years old, he met the other "germ theory father" in London. At the time, Pasteur was sixty-eight and partially paralyzed. The encounter was cordial, tense, and controversial. Both men were doing research on anthrax. After Pasteur presented the results of his research, Koch was judgmental. Neither man spoke the other's language. Letters were exchanged, and one of Pasteur's remarks was translated as a comment on "German arrogance." After the apparent insult, each man started criticizing the work of the other.

Once doctors had good microscopes, they learned how to categorize bacteria by drying and dyeing tissue and sputum that contained germs. Koch used special stains on infected human and bovine (cow) tuberculosis and identified the bacillus that caused the disease. "A plodding worker and a careful seeker of facts," Koch dazzled a group of colleagues on a Friday evening in 1882. He proved that the tubercle bacillus was transmissible and that it was the cause of TB in man.

Much as people today are investing our hopes and fortunes into a vaccine that will force our immune systems to reject the coronavirus, Koch tried to energize the (poorly understood) immune systems of people with tuberculosis. He isolated a glycerine extract of the TB bacillus and injected it into the skin of a person with an active TB infection. The fluid caused chills, fever, and an aggressive skin reaction. When it was instilled into infected guinea pigs, it seemed to "completely cure animals in the late stage of the disease."

Koch unveiled his new treatment when he addressed the crowd at a Berlin auditorium. "I have at last hit upon a substance which has the power of preventing the growth of the tubercle bacilli not only in a test tube but in the body of an animal." In the subsequent months, he began giving regular tuberculin injections to a number of patients with advanced disease who were in Berlin's Charité hospital.

Arthur Conan Doyle, the Scottish physician who created Sherlock Holmes, admired Koch and wanted to meet and hear the great man. On November 16th, he arrived in Berlin by train. When the British embassy was unable to get him a seat at one of Koch's demonstrations, he went to Koch's house. He knocked on the door, and the butler showed him into the living room. While Doyle was waiting, letters were dumped on a nearby desk and on the floor. Doyle would later characterize them as pleas for help from people with "sad broken lives and wearied hearts who were turning in hope to Berlin." The next day, Doyle visited the clinic where the infected were being treated. He saw people who were febrile, quite ill, and suffering as a result of the injections. Disappointed and dubious, he wrote about his visit and misgivings and returned to Scotland.

Koch's supply of his "remedy" was "scarce," but by the end of 1890 more than two thousand people with advanced disease had been treated. Most of the people who received tuberculin were not improved and only twenty-eight were cured.

Facing public scorn because his treatment failed, the now forty-seven-year-old Koch left his wife and married Hedwig, an eighteen-year-old art student who was "fascinated with his studies." He traveled to Egypt and wrote his eighteen-year-old lover, "If you love me I can put up with anything, even failure. Don't leave me now." When Koch inoculated himself with tuberculin, she volunteered to be injected too.

During his later years, his reputation now diminished, Koch traveled the globe with Hedwig and weighed in on various issues—often to his detriment. He, for example, didn't believe milk that contained bovine (cow) TB was harmful and opposed the pasteurization of milk. He also promoted the use of an arsenic-containing medicine to treat sleeping sickness. When he was sixty-seven, he had a massive heart attack.

During Koch's lifetime, many who had tuberculosis spent a year in a sanatorium. Breathing clean air and leading a healthy life helped some of them go into a remission. The first antibiotic that killed TB, streptomycin, was discovered in 1943. Like penicillin, it was being used by a soil organism to defend itself from the bacteria that surrounded it. Over time, streptomycin-resistant bacteria started to emerge. In 1953, it was joined by isoniazid (INH). The medication was a chemical that a Ph.D. student in Prague synthesized in 1912. It was probably sitting on a shelf somewhere when, in the 1940s, industry researchers decided to test hundreds of random chemicals on mice with tuberculosis. The third powerhouse, rifampin, became available in 1963. It was a chemical that was produced by a soil organism, and it was isolated and modified by Italian researchers.

By the 1950s and '60s, doctors were able to successfully treat most TB infections with a combination of the medications. Between 1954 and 1985, the number of infected people in the U.S. dropped from eighty thousand to twenty thousand, and experts predicted that within a few decades tuberculosis would disappear. Unfortunately, poverty, HIV, and bacterial resistance reversed the trend, and the incidence of TB started to rise.

At some point in their lives, one in four people alive today, will be infected by the cough of a person with tuberculosis. Ninety percent mount a cell-mediated immune response. Their body encases and imprisons the bug, but doesn't always kill it. Years later, the bacillus sometimes escapes, grows, and spreads. In 2019, ten million people worldwide developed an active infection and 1.4 million died.

Koch and Pasteur had challenged the belief that diseases were the result of some mysterious force in the miasma. They used live organisms to energize the immune system and with others demonstrated that germs cause disease and cleanliness matters.

Joseph Lister, the man who was called the father of modern anti-sepsis, began his medical studies in the mid-1800s. After he graduated he became the surgical apprentice of James Syme, "the greatest surgical teacher of the day." A humble Scotsman with an athletic build, Lister married Syme's eldest daughter, Agnes, and adopted her religion. Born a Quaker, he became a Scottish Episcopalian. A few years after he completed his training, Lister was the surgeon for the Glasgow infirmary. He noticed that half the people who had a limb amputated became septic and died, and "when the wounds were cleaned, some healed." "On the advice of a chemistry professor he read Pasteur's papers on putrefaction and he postulated that the process that caused fermentation was involved with wound sepsis." Based on his theory he started treating raw wounds with carbolic acid, a foul-smelling antiseptic that was used to clean sewers. His surgical infection rate dropped to 15 percent, and the Scots were impressed. Their doctors started cleaning and sterilizing the tools they used. Doctors in England weren't convinced until Lister went to London and operated on a fractured kneecap. He wired the bone together, closed the incision, and the wound didn't become infected. Over time he wrote articles and influenced his peers. At age fifty-six, he was named a baron.

Florence Nightingale took our awareness of cleanliness up a notch. Born in Florence, Italy, hence her name, she was the rebellious daughter of wealthy Brits who didn't want her to become a nurse. During the disastrous Crimean War between Britain and Russia, (1853–56), she worked at a small hospital in London. A world away, in Turkey, a muckraking reporter visiting the front lines stopped at a British military hospital. He found the conditions "appalling," which no doubt meant poor sanitation, gaping wounds, and bad smells. His newspaper articles detailed what he saw and his fellow countrymen were incensed. Then a high official made it possible for Florence to get involved, and she and thirty-eight other nurses sailed to Turkey.

At the military hospital in Scutari, sanitation was "neglected and infections were rampant." There was no clean linen. The clothes of the soldiers were swarming with bugs, lice, and fleas. The floors, walls, and ceilings were filthy, and rats were hiding under the beds. There were no soap, towels, or basins, and there were only fourteen baths for approximately two thousand soldiers. Nightingale purchased towels and provided clean shirts and plenty of soap. She brought food from England, scoured the kitchens, and set her nurses to cleaning up the hospital wards. Then a sanitary commission, set up by the British government, arrived and flushed out the sewers. She may not have had the drugs, blood, or modern-day "tools" that can turn an illness around, but she showed that diseased bodies have a remarkable ability to mend themselves. As Florence once wrote: "Sufferings were the result of too little fresh air, light, warmth, quiet, or cleanliness."

CHAPTER TWO

The Late-19th Century

H alfway through the 19th century, a dentist introduced anesthesia. A tombstone in a Boston cemetery marks the site of the "Inventor and Revealer of Inhalation Anesthesia: Before Whom, in All Time, Surgery was Agony; By Whom Pain in Surgery was Averted and Annulled; Since Whom, Science has Control of Pain."

Ether's effect was first demonstrated on a day in October 1846 in the amphitheater of Boston's Mass General Hospital. One of the nation's busiest, the hospital hosted up to two surgeries a week. They were performed in an operating room that was, in essence, a stage. It was surrounded by steep rows of seats where spectators sat and watched. The day in question, the doctor in charge, Dr. Warren, told the onlookers that "there is a gentleman who claims his inhalation will make a person insensitive to pain. I decided to permit him to perform his experiment."

The dentist who administered the anesthetic, Dr. Morton, was described as being "strikingly attractive and alternately optimistic and pessimistic." He arrived twenty-five minutes late, took out his narrow neck flask, and filled its bottom with two liquids: sulfuric ether and oil of orange. The second chemical was supposed to mask the ether odor.

The man who was about to undergo surgery inhaled "gas" through a mouthpiece, and in three to four minutes he became "insensible and fell into a deep sleep." He had a mass in his neck and the doctor quickly cut it out. When the operation ended, the man's surgeon spoke to the rapt onlookers. "Gentlemen, this is no humbug." People cheered, and the public took notice.

During the Civil War battle of Fredericksburg, Morton decided to help. When a wounded soldier was about to undergo a limb amputation Morton "prepared the man for the knife. He produced perfect anesthesia in an average time of three minutes."

See One, Do One, Teach One

Man has long known how to "suture lacerations and drain pus," but prior to 1858, when a gangrenous or damaged limb was amputated, resections were quick, bloody, and accompanied by loud shrieks. The fastest knife in England was wielded by Robert Liston, a Scottish surgeon who "operated with a knife gripped between his teeth." He could amputate a leg in two minutes. On one occasion, Liston inadvertently cut off the fingers of the person who held the patient down. The stump end of the man's hand and the raw end of the hip got infected and both men died.

In the 1800s, minor surgical procedures were carried out in homes. Doctors at a few hospitals operated on stages surrounded by tiered seats, and they thrilled observers with "spectacular public performances."

During the American Civil War (1861–1865), more than one hundred thousand injured soldiers were anesthetized before they underwent surgery. Thirty thousand limbs were removed, and three in four of the people who lost an extremity survived.

After the war, operating rooms were "quiet and to some observers the surgeries seemed tedious."

The first doctor who won a Nobel Prize for his surgical prowess, Theodore Kocher, was born, raised, and a lifelong resident of Bern, a centuries-old Swiss city that is encircled on three sides by the Aare River. He was once described as "a slight, rather cadaverous little man with a close cropped beard and very large, prominent upper teeth that made his smile rather ghostly." A colonel in the Swiss militia, he spent some of his non-medical time trying to get the weapons makers to create missiles that didn't deform and were "intentionally less lethal." Deeply religious and adamant about sterility, he told students whose patients had developed an infection to stand, beat their breasts, and say "I sinned."

In the days before we learned that iodide deficiency causes large thyroid goiters, a lot of people were walking around with a large lumpy gland in the middle of their neck. The condition is now rare because most countries require salt to contain iodide. In the U.S., the land of individual freedom, salt is often iodide free.

In the hands of most surgeons, removing the growths was a bloody, often lethal undertaking. Kocher, however, removed large goiters and cancer of the thyroid slowly, with little or no blood loss. The pace of his operations irritated some of the spectators who gathered for the show. By 1912, Kocher had resected approximately five thousand thyroid glands and his surgical mortality rate was less than 1 percent. Unfortunately, he didn't know that the thyroid gland produced an important hormone and he was unaware of the harm he caused when he totally removed it.

In 1882, a physician in Geneva reported on the changes that occurred to ten patients whose thyroids had been totally excised a few years earlier. Kocher heard the paper and asked a few people he had operated on to come in for an evaluation. One woman in particular bothered him. She "had changed from a pretty young girl to a small woman who was overweight, and slow of intellect and speech. She had lost hair and had a thick tongue." Realizing the lack of thyroid was the cause of her changes, Kocher "vowed to never again to remove the entire gland."

In 1895, Kocher wrote the first textbook of surgery. In it he explained how to make incisions in the skin without dividing important nerves or cutting open major arteries and veins. The book also has detailed instructions for removal of the thyroid gland, but it does not mention four yellowish glands that are the size of a grain of rice and are fastened, button-like, to the back wall of the thyroid. These are parathyroid glands, and they make a hormone that regulates the level of calcium in the blood, bones, and urine. When they are removed with the thyroid gland, people are prone to seizures and a slew of other problems.

In the 1880s, Reginald Fitz, a ninth-generation American, spent two post–medical school years as an observer in hospitals in Vienna, Paris, and London. When he returned, he became a pathologist. At one point he told a group of Harvard surgeons that "most inflammatory disease of the right lower quadrant begins in the appendix." He recommended surgical excision when the appendage was inflamed. During the next few years, a few Boston surgeons tried to remove the fingerlike projections but had trouble seeing what they were doing. Their incisions ran down the middle of the abdomen, and the appendix was in the right lower corner. Surgeons also removed stones from inflamed gallbladders. They didn't start cutting out the actual gallbladder until two German surgeons wrote an article explaining how to successfully perform the operation.

Although operations could now be performed without pain, surgeons who encountered a person with worrisome abdominal distress or a stab wound often didn't know what to do next. Some doctors spent months or years as an apprentice of an established surgeon, and some apparently winged it. No one really knows how the men who performed surgery in the 1800s learned their craft.

Drinking Water and Diarrhea

In the early 1800s people knew that some water made them ill and they "determined the quality of water based on its taste." Children drank a brew made from boiled water and adults tended to imbibe ale. In 1854 an epidemic of diarrhea broke out in the Soho district of London. Many became dehydrated and more than six hundred people died. At the time, most households didn't have piped-in water. People used communal pumps and stored their fecal matter in nearby cesspools. Sewage was dumped in the Thames, and the river smelled.

A London physician named John Snow saw his patients suffering and suspected an intestinal poison had contaminated the local water. A vegetarian and teetotaler, Snow mapped the location of the people who contracted the illness and convinced town officials that the Broad Street pump was the source of the epidemic. They removed the pump handle and the diarrhea "almost immediately" trickled to a stop. Snow was, of course, disbelieved and denounced by many.

No one knows what caused the outbreak, but many suspect it was cholera. The lethal disease emerged from the Ganges Delta of India in the 1800s and caused a series of worldwide epidemics. Infected people developed severe diarrhea and became dehydrated. Many died. The disease was spread by ships trading goods in Thailand, China, and Japan. In 1831, a vessel whose sailors had the illness docked in the English town of Sunderland. After they disembarked, the infection spread through the U.K. and killed thousands.

Twenty-nine years after the pump handle was removed, the German researcher Robert Koch traveled to Egypt during a cholera outbreak. He cultured and examined the stools of the people who had the infection and he identified the responsible germ. It was a "bacillus and it was bent like a comma."

In the late 1800s, people learned that waterborne outbreaks of typhoid fever, dysentery, and cholera were caused by bacteria and that chlorine killed the germs. In 1905, a municipal water sand filter failed in a city in England, there was a typhoid outbreak, and the city started adding chlorine to their municipal water. Three years later, Jersey City, New Jersey, joined them. Chlorine is currently added to "over 98 percent of all U.S. municipal water," and outbreaks of typhoid fever have been virtually eliminated. (Chlorine does not kill parasites like *Giardia* and *Cryptosporidium*.)

In the last century, we've learned a lot about diarrheal diseases and their treatment, but half the world's population still lives in tightly packed dwellings and don't have pipes that bring water in or fecal waste out. When people defecate outside (because the feces smell), their waste enters streams that flow into the rivers that are the source of drinking water. According to the CDC, one in nine childhood deaths worldwide is the result of diarrhea. The culprits responsible include bacteria like *E. coli*, salmonella, shigella, and cholera.

When I was a freshman med student, a biochemistry professor who looked like the movie star Robert Wagner spent ten years of his life proving that the water we drink isn't merely passively absorbed. Salt and water are sometimes pushed into our body by a biochemical "pump" that is fueled by glucose.

I took notes when the professor lectured and I passed the exam, but I didn't understand why a person would spend years of his life learning how salt and water move across the small bowel membrane. I later learned that when the cholera bacteria infect the small bowel, the organism forces a large amount of liquid to leak into the lumen of the intestines. Infected people develop watery diarrhea, don't absorb much fluid, and often die of dehydration.

Our lecturer, Robert Crane, showed that if a teaspoon of salt and two tablespoons of sugar are added to a liter of clean water, a biochemical reaction powers the liquid across the membrane of the small intestine, and people can be rehydrated. The World Health Organization distributes a version of Crane's solution, and drugstores in the U.S. sell it as Pedialyte. The discovery of the sodium and sugar "co-transport" mechanism has saved more lives than most of the expensive medications advertised on television.

Crane was born in New Jersey and wasn't interested in science before he entered college. During the Second World War, he was a deck officer on a navy destroyer that "took a bomb" at the Battle of Leyte Gulf. His vessel was later part of the fleet of ships that carried Douglas MacArthur back to the Philippines in 1944. After the war, he earned a Ph.D. in chemistry at Harvard, married a biochemist, and started working on his pump theory. In 1978, he was awarded the Nobel Prize.

In 1991, a massive hurricane blew across Hispaniola, the Caribbean island that is shared by the Dominican Republic and Haiti. At the time, Haitians hadn't witnessed a case of cholera for more than a hundred years. The U.N. brought in relief workers from Bangladesh and they lived with the people. Seventy percent of the island's households had either rudimentary toilets or none at all. A few months later, the water people drank contained the strain of cholera bacteria that was common in Bangladesh. Approximately 665,000 Haitians developed severe diarrhea and more than 8,000 died. In 2006, fifteen years after the epidemic started, a nonprofit called Sustainable Organic Integrated Livelihoods (SOIL) started supplying compostable toilets and trying to help solve the island's waste problem.

A Number of Parasitic Diseases and Their Vectors

In the mid-1800s and 1900s, many believed that the chills, fever, aching, and exhaustion of malaria were caused by a vengeful god or toxic fumes rising from a swamp. (Malaria is the Italian word for bad air.) Then, in 1880, a French military doctor sighted parasites that were attacking a sick person's red blood cells and making people ill. The physician who made the discovery, Alphonse Laveran, spent his early career looking for infectious agents in the soil and water of Italian marshes. He found and described a number of protozoa and bacteria that lived in the marsh-lands, and he published a "Treatise on Military Diseases and Epidemics." During the 1870 war between France and Germany, Laveran was sent to the front, and he was captured and imprisoned. After the conflict ended, he was freed and stationed in Bone, Algeria, a colonial city on the Mediterranean coast. As the officer in charge of the French military hospital, he encountered wards full of sick young recruits. They suffered fever, joint pain, and exhaustion—malaria. Some died. Quinine, the dried bark of the South American cinchona tree, was being used in the Americas to fight the disease, but I don't know how the French in Algeria treated the problem.

Laveran let the nurses take care of the soldiers, and he performed autopsies. By the late 1800s, microscopes that could visualize very small objects were widely available. Laveran pricked the fingers of the sick, smeared their blood on slides, and studied the preparations with a micro-scope. At some point, Laveran noticed that the red blood cells of people with malaria contained small black granules or pigments. The fragments were absent in people who did not have the disease. One November morning in 1880, while studying blood drawn from a sick soldier, Laveran saw "filiform elements on the edges of red cells that moved with great vivacity." Their motility convinced him that he had discovered the agent causing malaria and that it was a protozoan parasite.

When he attended a medical conference and presented his findings to fellow physicians, his discovery "was received with much skepticism." At the time, another investigator described a bacillus in the soil and water that he thought caused malaria. Laveran's "agent" did not resemble any-thing that looked like a familiar pathogen . . . and many observers who didn't know how to classify the strange object found it simpler to doubt its existence. By 1884, a few Italian researchers who looked through micro-scopes at blood taken from people with malaria also observed "amoeboid movement of strange organisms." Their discovery validated Laveran's, but many of the day's doctors continued to believe that malaria, like other tropical diseases, was the result of a country's climate, ground water, humidity, and vegetation.

In 1876, mankind got a step closer to understanding the role mosquitoes play in transmitting disease. That year a Scottish doctor named Patrick Manson returned to China with his new wife and a powerful microscope. As a youth, Manson had planned to become an engineer, but it was physically too difficult and he "trained in medicine." After he graduated, he briefly worked in a lunatic institution. Then he embarked on a three-month ocean voyage and joined his doctor brother on the Chinese island called Formosa. Manson spent the next nine years providing medical care to the Brits who worked there. He also got involved in the plight of the impoverished locals. Working at the Protestant mission in the nearby village of Amoy, Manson introduced smallpox vaccinations and taught a few locals how to perform minor operations. One day he cared for a nineteen-year-old who drank arsenic, vomited, and survived. The man had tried to kill himself because he didn't want to live in his deformed, swollen body. He had a disease called lymphatic filiariasis. The condition causes arms, legs, and scrotums to swell, and people who have the condition often feel like outcasts. We now know the disease is caused by microscopic, threadlike worms. They plug the lymphatic vessels. the tiny channels that transport fluid and infection-fighting white cells toward the head. They empty into the veins under the collarbone. As the fluid flows, it is filtered by lymph nodes.

After doctoring for nine years in the Far East, Manson returned to England and married. While in London, he read the little that was known about lymphatic fluid. Then he returned to China with his new wife and a good microscope. During the next 11 years, Manson carefully examined the blood of people with elephantiasis. He found tiny creatures, filariae, were floating in it. Suspecting the disease might have been spread by mosquitoes, Manson learned how to dissect the tiny insects and examine their intestines. Some of the mosquitoes he checked contained threadlike parasites. Over time, Manson observed organisms that grew and changed shape. Much as a butterfly spends part of its life as a caterpillar, the creatures that cause elephantiasis spend part of their life cycle in insects and part of it in man.

After twenty years in China, Manson published his findings and returned to England. Needing money, he set up a practice in London and cared for Brits who returned from the tropics with a variety of ailments.

One day he met Richard Ross, a vacationing doctor who worked in India. Ross was born in India, and had lived there briefly as a child. He never really wanted to become a doctor. He instead planned to write books or to work in mathematics, art, or music. His father, however, told Ross he had to go to medical school and he went. Never a good student, Ross graduated, was certified, and decided to partake in the good life Brits enjoyed in India. Barely passing the Indian medical service exam, Ross was not held in esteem. He did his job and saw a few patients when he wasn't playing golf, hunting, or writing his novel. During his first furlough in London, he got a master's in public health. He hoped it would help him move up the ranks. He also married and brought his wife back to India. Years later, he returned to London for another furlough. By that time, he had read about Laveran's assertion that he saw parasites in the blood of people with malaria. Like many doctors of the time, Ross was skeptical. He had examined a lot of smears of blood taken from people with malaria and he had never seen one of the so-called parasites. At one point, he visited Manson's office and asked the tropical expert what he thought. Manson walked to a microscope and within a few minutes taught Ross how to see the malarial forms. The two doctors became friends and Manson told Ross that he believed malaria was spread by mosquitoes. He couldn't prove it because he was stuck in London. As they talked, Manson convinced Ross to do the research in India.

Ross returned to British Raj, and tried to prove Manson's theory, but it wasn't easy. Ross had to pay people before they would allow him to draw their blood, and he sometimes didn't have enough money. The Indian bureaucracy periodically moved Ross from one location to another. At one point he was working with the wrong mosquito. During those years, Manson advised, coached, and encouraged Ross. The two men exchanged over 100 letters. Each note didn't arrive until four months after it was written. Eventually, Ross learned how to dissect and examine mosquitoes.

After three years, he found a pigmented cell in the stomach wall of a mosquito. It turned out to be a developing parasite. Then Ross found an early parasite in an insect's salivary glands. He ultimately had to use birds to prove mosquitoes transmitted the parasite.

When Ross's work was published, some remained skeptical, and Manson staged an experiment to convince nonbelievers. He shipped a number of infected mosquitoes from Rome to the middle of London. The trip by ship took forty-eight hours and some of the mosquitoes died en route. When they arrived, Manson allowed one of the mosquitoes to infect his son, a doctor. The young man got sick, but was cured with quinine.

For his efforts, Ross received a Nobel Prize. While Ross was making his discovery, a few Italians were also proving mosquitoes were responsible for transmitting malaria, but they did not get part of the prize.

One hundred twenty years later, malaria doesn't affect people who live in countries that are able to control the vectors that spread it, but forty percent of the world remains at risk. According to Nicholas White, the Brit who is a professor of tropical medicine in Thailand, "Malaria is currently the most important parasitic disease of humans." It has a huge humanitarian and economic impact, mainly in Africa, and it causes more than one thousand deaths each day. Most who die are children.

Centuries before Western man learned what causes malaria and how it's transmitted, natives in South America were using the quinine-rich bark of the cinchona tree to treat the "shivers."

In the 1930s, Hans Andersag, a Tyrolian-born researcher who worked at the Bayer facility in western Germany, deconstructed quinine and developed chloroquine, a drug that became an extremely effective anti-malarial medication. During the mid-20th century, it was used all over the world. Brazil and a few other countries added the drug to salt. The United States Agency for International Development (USAID), the "humanitarian" agency created by U.S. president Kennedy, distributed over a hundred million chloroquine pills.

In the 1960s, the malaria parasites gradually became resistant to our drugs and the disease started reemerging.

In 1966 Mao Zedong launched the Cultural Revolution. He closed the schools and gangs of young Chinese—the Red Guard—humiliated, assaulted and sometimes killed Mao's enemies. That year the war between America and the North Vietnamese heated up. Viet Cong soldiers lived in the jungles and many developed chloroquine-resistant malaria and became ill or died. The leaders of North Viet Nam asked China for help. Under Mao's direction, a large number of researchers sought an herbal solution.

"By 1972, researchers for the U.S. had screened more than 214,000 compounds, and Chinese military researchers had screened thousands more. None of the drugs were effective. In 1969, Chinese leaders sought help from the Academy of Traditional Chinese Medicine. By that time, the Cultural Revolution had sidelined the academy's experienced experts, and the institution ended up putting a junior researcher in charge. Her name was Tu Youyou. Tu was twenty-nine years old and was married. Her eldest daughter was four years old, her youngest daughter was one, and her husband had been sent to a 'training campus.' To perform the research she had to leave home. She left her youngest daughter with her parents in Ningbo and sent her eldest to live with her teacher's family. Tu wasn't able to visit her parents for the next three years, and when she did the then seven-year-old child hid behind her teacher.

"As part of her research, Tu read Chinese books that were written thousands of years earlier, and she collected more than two thousand herbal recipes. Her group spent years testing the extracts of more than a hundred herbs on rodent malaria, but the drugs kept failing. Then Tu read about an herb that showed some anti-malarial effects that were inconsistent and hard to reproduce. An ancient herbalist wrote about immersing a handful of Qinghao (now called artemisinin) in two liters of water. He wrung out the juice and drank it. Tu knew that most herbs were boiled in water, and she wondered if heating the herbs during extraction might destroy the active components. Deciding to extract sweet wormwood stems and leaves at a reduced temperature, she used water, ethanol, and ethyl ether. At the time, most pharmaceutical workshops were shut down by the great Cultural Revolution, so to extract the herbs her group had to use household vats.

"On October 4, 1971, Tu's group observed that sample number 191 of the Qinghao ethyl ether extract effectively inhibited malaria parasites in rodent malaria. Tu reported the findings at a national meeting held in Nanjing, but shortly thereafter her research was put on hold. There were reports of animal toxicity. Undaunted, Tu got permission to take the extracts voluntarily, and two team members joined her. They were observed in hospital, and no side effects were observed." The herb she discovered is called artemisinin, and over the years it has been widely used to treat and prevent malaria and save lives. Tu became a professor at the Academy of Traditional Chinese Medicine, lived with her husband in an old apartment building in Beijing and, in general, was "almost completely forgotten by people." The World Health Organization didn't begin recommending "artemisinin-based combination therapy (ACT) for the treatment of uncomplicated falciparum malaria" until she was eighty-two years old. In 2015 Tu was awarded a Nobel Prize.

In the 21st century, the Swiss company Novartis packaged and sold Coartem. The medicine is part artemisinin and part a weaker drug. The combination of two or more medicines lessens the likelihood that a resistant parasite will emerge. In 2009, the Gates foundation started contributing to the cost of drugs that were used by people in Africa.

In the late 1800s, the French started building a canal between the Atlantic and Pacific Oceans. At the time, few knew the mosquito transmitted malaria and another common and lethal tropical disease called yellow fever. The construction was started eleven years after the Suez Canal opened, and promoters claimed an ocean-to-ocean passageway in Panama wouldn't be very expensive or difficult. They raised money from more than two hundred thousand investors, hired twenty thousand workers from the West Indies, and put a Frenchman named Jules Dingler in charge. When he went to Panama, Dingler was accompanied by his family and a few stallions. He is credited with having said that "only drunkards and the dissipated contract yellow fever and die." During his early years on the job, Dingler's son, daughter, and wife each contracted yellow fever. They all died and Dingler, a man "broken in mind and body," shot his stallions and returned to France.

Unlike the flat Egyptian landscape that bordered Suez, the Panama terrain turned out to be challenging. Between 1881 and 1889, more than twenty thousand laborers who toiled in the thick rain forests died. In

1889, the company declared bankruptcy and the French abandoned the project.

In 1900, an American military doctor was sent to Cuba to investigate a yellow fever outbreak. At the time, the island was occupied by U.S. soldiers. America had won a war with Spain in 1898 and had freed the island from the European colonizer. Cuba was scheduled to become independent in 1902. During the U.S. occupation, some of the American soldiers who were stationed on the island developed headaches, fever, bled easily, and a few died.

When Major Walter Reed arrived on the island of rum and cigars, he met with local doctors who apparently thought the illness was caused by mysterious "fomites." Reed was a military physician who had spent sixteen years in the arid American West. Before coming to Cuba, he was a professor at the army medical school. While there, he read articles written by a Cuban ophthalmologist named Carlos Finlay. In 1881, Finlay used a mosquito to infect a twenty-two-year-old Spanish soldier and had "hypothesized" that the insect transmitted yellow fever. His presumption was "ridiculed by his colleagues," but Reed thought it was sound. The doctors who cared for the sick soldiers were willing to check out Reed's theory, though I doubt they thought it made sense. Volunteers, some infected, some well, were housed in a barracks. Mosquitoes were collected and allowed to feed on the blood of a sick soldier. Then they were brought to a healthy young man. The insects inserted the tips of their straw-like mouths into the person's skin and sucked their blood.

One of the apparently skeptical doctors who conducted the investigation "submitted to the bite of a creature that had fed on an infected soldier." He joked, "If there is anything to the mosquito theory, I should have had a good dose." A week later, his life was in the balance for three days.

A fellow doctor who brought the mosquitoes from one soldier to another noticed a bug on his hand. He allowed it to suck his blood. A week later, he became febrile, delirious, started vomiting, had seizures, and died. As one of the infected who survived put it: "Such is yellow fever."

We now believe yellow fever came from the rain forests of Africa and was brought to the New World by slave traders. The disease is caused by a virus, and there is an effective vaccine. Each year it causes two hundred thousand infections and thirty thousand deaths in central Africa and South America.

In 1904, when the U.S. started building the Panama Canal, the people in charge knew that the mosquito transmitted malaria and yellow fever. Engineers "drained pools of water, cut all the brush and grass near villages, and constructed drainage ditches." Larvae were oiled and killed with an insecticide. Screens were placed on windows and doors. "Collectors were hired to gather any adult mosquitoes that remained in the houses and tents during the daytime." It took ten years but the U.S. successfully built the canal that linked the two oceans.

The world's second most prevalent parasite, bilharzia (shistosomiasis), affects two hundred million people in about eighty countries. Ninety percent of the affected live in sub-Saharan Africa, but the parasites also infect people who live in South America and a number of Asian countries. The creatures spend part of their life cycle in snails that live in fresh water ponds. When they exit their host, they enter the water, and penetrate the skin of an animal or person who is wading. Inside a body the parasites make their way to the liver or urinary tract, lay eggs, and can cause significant damage.

In 1960, I spent the med school summer at a U.S. Army parasitology lab in Puerto Rico. The lab was located on a bluff overlooking the balmy Caribbean. While there, a fellow student and I drove to an inland body of water and put on protective hip boots. Like the 1915 researchers who worked out the parasite's life cycle, we trudged into the ponds where the snails live. We scooped the snails up with a net, carried them back to the lab, and placed them in sunlight. The snails discharged, "shed," miniscule worms. We gathered them, placed them in test tubes, and dipped the tails of mice into the liquid. Weeks later we studied the infected mice. At the end of the summer, I returned to school and as a physician I never encountered a live person who has the disease.

In the 1970s, Merck scientists found that one of their chemicals, praziquantel, effectively and relatively safely kills most of the shistosomiasis parasites that are camped in a person's liver or in other parts of their body. They marketed the drug for animals and later for humans. When it first became available, the drug was too expensive for most of the more than two hundred million people who were chronically infected. At the time, Korea didn't allow manufacturers to patent a drug. They were, however, allowed to patent their process for making it. Chemists working for the Korean company Shin Poong learned how to produce praziquantel using an alternate chemical approach, and they produced and sold the medication at a more affordable price. In 1995, the World Trade Organization (WTO) was founded. They required anyone who wanted to trade with another WTO country to patent the actual drug for twenty years. The rule presumably helped wealthy pharmaceutical companies protect their "intellectual property." I don't know what impact the rules had on the ability of the planet's poorest to access needed medications.

Over the decades, praziquantel's original patent expired but the World Health Organization decided to target the parasite. Merck donated more than 500 million pills to the organization. When the pills are given to infected children the eggs in their bodies are destroyed and the kids are healthier and don't transmit the disease.

In 2017 ticks were responsible for most of the sixty thousand annually reported vector-borne diseases in the U.S. The most common infection that ticks spread is caused by a bacterium that's technically a spirochete. Called Lyme disease, it is named for the town in Connecticut where it was identified. In the mid-1970s, Mrs. Murray, an artist, was living in Lyme in a house near a picture-postcard rural country road. She developed rashes, painful swollen joints, numbness, and weakness. Over the next few years, she got worse. She was hospitalized three times, was unable to paint, and some thought she was a hypochondriac. Doctors started paying more attention to her problem when her son developed joint pain and couldn't smile. He had Bell's palsy and his facial muscles didn't work. Another son developed a rash behind his knee and others in the town got ill. It took a while before a young Yale physician and others discovered that some ticks carry a bacterium—a spirochete that they inject into a person's body. If it's not detected and treated early, the bacteria can spread to the heart and brain and cause chronic problems.

When one of the eight-legged arachnids is infected with babesiosis and bites, the parasite enters the body. It invades red cells, causes chills and fever, and can be lethal. During the last century, we've become aware of one, then another, of the diseases that are transmitted by the bites of ticks. The creatures hide in the long grass we hike through, the nearby woodlands, and in the fur of one of our four-legged companions.

Intestinal Parasites—DES

Prior to the 1800s, people knew that strange wormlike life forms live in many intestines. They were easily seen in the feces of an infected person or animal and were so common that in the 1700s many believed they were made in a person's body or by spontaneous generation.

In many parts of the world, intestinal parasites are still quite common. When people defecate in the open (and half a billion people in parts of the world still do), bacteria and parasites are deposited in the dirt and they get into our rivers. They spend part of their cycle of life in cows or fish and can enter our bodies when we don't cook the food we eat. Some, like *Ascaris*, and hookworms are able to penetrate the soles of our feet. We are at risk when we walk barefoot on dirt where someone defecated.

One day when my grandson was three or four, he passed a worm. He had never lived in a third-world country and we still don't know how or where he acquired the creature. My daughter noticed the "whatever-it-was" in the potty, checked the Internet, identified the parasite, and put it in a jar. When she showed it to her doctor, he was taken aback—amazed. His nurse, on the other hand, merely shrugged. No big deal. She had grown up in a village in the Philippines.

I first learned about parasites when I was a medical student. As a freshman, I attended a weekly class and was introduced to pictures of a large number of bizarre microscopic creatures. I don't recall many of the details that I crammed into my brain for the final exam, but I'll never forget the take-home message. It was written on the chalkboard by the good-natured Japanese professor and is probably the only thing that most of the giggling medical students remember to this day. DES. Don't Eat Shit.

The chief drug that's used to kill intestinal parasites recently became a very expensive commodity. Patented in 1975 by Smith Kline, it was created by Robert J. Gyurik and Vassilios J. Theodorides, after Vassilios read an article, had a sudden insight, and sketched the chemical structure of the future medication.

The drug's inventor was raised in a small Greek village near the Macedonian border. When he talks about his youth, his tale begins on a morning in 1941. Four hundred German soldiers surrounded his town and marched its sixteen hundred occupants to the village center. Vassilios was ten years old at the time. The soldiers told everyone to bring out their guns. Then they searched the houses and found shotguns in two homes. The owners of the weapons were tied to a tree and publicly executed. Later in the day, a German soldier searching Vassilios's house saw a shotgun behind a door. He motioned to Vassilios to hide it, and when the soldier left, the Greek boy threw the gun in the bushes.

The school Vassilios attended was in a village an hour and a half walk away. He was in class the day in 1947 when Communist soldiers burned his hometown and killed forty-eight people. Some were relatives.

After he finished high school, he planned to become a mathematician. He visited the university office and asked where the mathematics department was. The clerk asked what was wrong with the veterinary school and Vassilios decided to give it a shot. As a veterinary student, he developed an interest in research and decided he needed a Ph.D.. His future wife's family had immigrated to Boston and Vassilios decided to follow her. He came to the U.S., married the girl, earned a Ph.D., and worked for Pfizer in Terra Haute, Indiana, for two years. There were only two Greek families in town, and his wife was unhappy so they moved to Pennsylvania, and he got a research job at Smith Kline & French laboratories.

A few years later, he read an article and had an "aha" moment. Unexplainably, he somehow "knew" the steps he would have to take to create the chemical that became albendazole. (Quoting Pasteur, he explained, "God helps the minds that are prepared.") The medication he created was introduced in 1977 and was initially only given to animals in Australia and New Zealand. It was not approved for people in the U.S. because someone at the FDA decided it was carcinogenic. Vassilios met with scientists at the agency. He showed them they were using an incorrect mathematical approach when they evaluated the drug's potential carcinogenicity. The officials at the FDA agreed and humans started using it in 1982.

Twenty eight years later, the company that owned the drug had merged with two other pharmaceutical giants and was called GlaxoSmithKline (GSK). It had offices in more than one hundred countries, and its London headquarters was located near the spot where Julius Caesar purportedly crossed the Thames River during his 54 BC invasion of Britain. By October 2010, albendazole had become a financial loser, and the company dumped/sold it and the U.S. marketing rights. They were picked up by Amedra Pharmaceuticals, a small American company. The details of the deal were not disclosed (or at least I couldn't find them on the web). As part of the agreement, GSK agreed to continue manufacturing the medication for Amedra in the short run. They also renewed their pledge to the World Health Organization. They would continue to give the organization 600 million tablets per year as their contribution to the struggle to free the world from lymphatic filiariasis. Filariasis is a parasitic condition that affects 120 million people in eighty countries, and I discussed it earlier in the chapter.

The year after Amedra bought albendazole, the generic drug company Teva stopped manufacturing the drug's only U.S. competitor, mebendazole (brand name Vermox). Amedra became the only U.S. player in the intestinal parasite business and the drug had the potential to become a high-priced commodity. Impax purchased Amedra for $700 million and dramatically increased the amount they charged for the medication. In late 2010, the average wholesale price was about "$6 per typical daily dose," and by 2013 it had jumped to almost $120.

In 2020, economist Paul Krugman commented that many Americans believe that "unrestricted profit maximization by businesses is the recipe for a good society." His sarcastic observation was a jab at the way our leaders allow our intellectual property and patent laws to be abused and our taxpayers to be exploited.

In 2008, Medicaid spent less than $100,000 per year on albendazole, and in 2013 the U.S. shelled out $7.5 million for the formerly inexpensive drug. Doctors in this country are prescribing it more often because the CDC (Centers for Disease Control) thinks we should presume that refugees that come here from poor countries have parasites in their intestines, and we should treat them. In 2017, according to an NPR report, the treatment of hookworm cost $400 in the U.S. and $0.04 in Tanzania.

CHAPTER THREE

The 1900s

In 1900, 76 million people lived in the U.S. and two hundred thousand miles of railroad tracks crisscrossed the continent. People traveled in horse-drawn carriages and wagons on narrow dirt and gravel roads, and trains dominated commerce. The modern internal combustion engine was fifteen years old, and 1,575 electric and 900 gasoline-powered vehicles were produced each year.

Electricity was new and scarce. Thomas Edison's incandescent light bulb was twenty years old and an American city, Cleveland, had started using electric lamps to illuminate some of its streets.

Sixty-seven years earlier, drinking water was pumped into the White House from a nearby reservoir; Chicago's "comprehensive sewer system" had been up and running for fifteen years, and a revolutionary toilet made by Thomas Crapper was nine years old.

In 1903, the Wright brothers kept a powered plane aloft for fifty-nine seconds, and in 1844 William Morse was the first to wire a message from one city to another.

At the turn of the century, house calls were common. People who had suffered heart attacks, strokes, or major trauma were often cared for at home. There were no antibiotics or blood transfusions, but adequate anesthesia made surgery painless, and doctors who dared were cutting abdomens open, peering inside, and learning what to do next. Small-town doctors who were usually respected or admired were generally only paid if there was enough money for the grocer, landlord, shopkeeper, tax collector, and plumber.

CHAPTER FOUR

Insulin

The Canadians who first isolated insulin obtained the American patents for the hormone and sold them to the University of Toronto for $1.00 each. Currently the hormone is produced by three companies. It costs much more in the U.S. than it does on the other side of America's northern and southern borders. Many who live near Mexico or Canada get their insulin by walking across, buying what they need, and "smuggling" it into the country.

Insulin is a hormone that is created in one of the clusters of endocrine cells that are scattered throughout the pancreas. After it enters a person's blood, it attaches to the wall of a fat or muscle cell, and convinces the cubicle to let glucose in. When sugar can't enter cells, the level of glucose in the blood rises.

There's a form of the disease where a person can't make any insulin. In those who have the condition, the immune system (for some as yet unknown reason) has attacked and destroyed their "beta" insulin-producing cells.

The pancreas is located in the upper abdomen, is shaped like a fish, and makes digestive enzymes that pour into the small intestine through the pancreatic duct. When the juice reaches the intestine, the enzymes are activated and they break food into small, absorbable pieces and divide the bonds that link the amino acids together to form proteins.

In 1920, scientists knew insulin was made in the pancreas but they didn't know how to extract it without activating the digestive enzymes and destroying it. When an animal was killed (for food), the enzymes became functional. The pancreas digested itself and the insulin it made. It was also known that tying off the duct that connects the pancreas to the small bowel activated a creature's digestive enzymes. The pancreas destroyed itself, but for some reason the creature didn't develop diabetes.

A Canadian surgeon, Frederick Banting, and his student Charles Best decided to check it out. Banting had been a battalion medical officer during the First World War and had an arm that had been damaged by shrapnel. After the war, he had trouble finding work and needed the research job. His assistant, Best, had just graduated from college.

The twosome tied off a dog's pancreatic ducts and managed to keep the dog alive long enough for the enzymes to become activated and destroy the pancreas. Then they removed the organ. They chopped it into pieces, mixed it with saline, and filtered an extract. They injected the extract into a diabetic dog and the animal's blood sugar dropped dramatically. As they hoped, the activated enzymes had broken down the cells that produced digestive secretions but hadn't destroyed the cells that made insulin. Months later, they repeated the experiment using a cow. This time a chemist named James Collip purified the extract. They injected the fluid into a fourteen-year-old with high blood sugar and his glucose level dropped dramatically.

The discovery was momentous. A fatal disease could be controlled. As mentioned earlier, in 1923, the Canadians who first isolated insulin obtained the American patents for the hormone and tried to make sure it remained affordable. They sold the American patents to the University of Toronto for $1.00 each. Frederick Banting declared, "Insulin belongs to the world, not to me."

The research director at Eli Lilly of Indianapolis, Indiana, read about insulin's discovery, came to Toronto, and obtained the right to develop and produce the hormone in the U.S. About the same time, a Danish physiologist named August Krogh and a few colleagues started making and selling the hormone in Europe. (The company they founded is now Novo Nordisk.) In Canada, Connaught Labs, with the help of Charles Best, figured out how to produce 250,000 units of insulin weekly. They were able to manufacture insulin and drop its price from $0.05 to $0.02 per unit.

The decision to award the 1923 Nobel Prize for the breakthrough to Fred Banting and J. R. Macleod was "much debated." Macleod was the chairman of the physiology department in Toronto and apparently approved the research, but some on the Nobel committee found it "difficult to evaluate Macleod's contribution." Charles Best was never nominated and that bothered him. He had done most of the work, and as time passed he "developed a deep psychological need to be recognized as a discoverer of insulin." During the subsequent decades, he and Banting worked together and "Banting developed an intense dislike of Best." In 1940, Best was invited to come to London. When he suddenly backed out, Banting, now the head of the department, decided he would go. Before boarding the plane and setting out for the distant world, before flying to war-torn London, Banting's last words were: "If they ever give that chair of mine to that son of a bitch Best, I'll roll over in my grave." His plane crashed in Newfoundland, and he died.

During the next half century, people used the insulin that came from cows and pigs. It worked as well as human insulin and it was chemically quite similar.

Human insulin is a protein that contains fifty-one amino acids. Porcine and cow insulin differ from the human hormone by one and three amino acids. At times impurities caused reactions, and some people developed antibodies to the animal products. Contaminants were initially a problem, but in the 1970s manufacturing improvements led to insulin that was thought to be very pure.

In the late 1970s, researchers at Harvard and in California, separately and competitively, used genetic engineering and tried to make synthetic human insulin. In 1980, six researchers in California were the first to fabricate the hormone.

The scientists were employees of Genentech, a company whose seeds were planted in 1973. Back then a biochemist from UCSF (University of California in San Francisco) and a physician from Stanford University met at a conference in Hawaii. At the end of a long day, they "took in the balmy evening air as they strolled and talked."

One of them, Herbert Boyer, "blue jean clad, with a cherubic face; outwardly relaxed and unassuming," had grown up in a small railroad town near Pittsburgh and, as a college student, had at times hitchhiked to classes at a nearby college. Majoring in biology and chemistry, he was "really taken with the Watson-Crick structure of DNA." He earned his Ph.D. in bacteriology, and at age thirty-seven he was a researcher at the University of California in San Francisco. While he was there, one of his graduate students isolated an important enzyme. It sliced DNA at a specific position. The raw exposed nucleotide ends were sticky and other lengths of DNA could be attached.

The other man was Stan Cohen, a thirty-six-year-old "trim, bald, bearded" Stanford hematologist. He was studying circular collections of DNA. Called plasmids, they were normally present in the cytoplasm of bacteria, and are sometimes passed from one bacterium to another by direct contact. It's one way antibiotic resistance is spread. The two investigators wondered if it was possible to use Boyer's enzyme to hook a gene (a known chain of DNA) onto the sticky ends of a plasmid's DNA. If the plasmid was then inserted into the cytoplasm of a bacterium, would the transformed plasmid survive and clone itself? Would the gene direct the bacteria to make a specific protein?

It took a few months to do the research, but the following March they tested their idea and it worked. In November 1974, both medical schools filed a patent application, and the academic world debated the potential hazards of genetic engineering.

Over the next few years, surviving on money gathered by a venture capitalist named Bob Swanson, Boyer formed a company and called it Genentech. In its early years, the company made somatostatin. It's a peptide that reduces secretory diarrhea and blocks the action of some hormones like insulin and growth hormone. It was not a big moneymaker.

Genentech then produced genetically engineered human insulin. The company's scientists added genes to plasmids and inserted the plasmids into the cytoplasm of the bacteria *E. coli*. The genes took control of the bacteria's protein-producing machinery. The *E. coli* manufactured two protein chains and they chemically combined and formed insulin. The liquid was isolated, purified, and tested. It was chemically identical to human insulin and it didn't cause any allergic reactions.

In 1978, Genentech leased a ten thousand–square-foot section of an airfreight warehouse near the San Francisco airport. By 1980, the company's technology was up and running and Genentech had a public stock offering. It was wildly successful and Swanson, one of the founders, called gene cloning "the cornerstone of a future billion dollar business."

In 1978, a few years before Genentech started marketing human insulin, one of the company's scientists visited an Indiana factory that was extracting the hormone from pigs and cattle. He learned that the manufacturer, Eli Lilly, was using the pancreases of 56 million animals each year. It took eight thousand pounds of glands and twenty-four thousand animals to make a pound of insulin.

After human insulin was introduced, it took a few years before people who had been injecting animal insulin into their bodies were ready to switch. But manufacturers stopped making the animal hormone, and twenty-five years later, most of the people who need insulin are using the human protein.

In 1923, when he gave away the hormone's patent, Banting had proclaimed, "Insulin belongs to the world, not to me," but in the 21st century some believed insulin should be treated as a valuable commodity.

In 2007, the American list price for a five-pack of a brand of insulin made by Eli Lilly (according to Vice News) was $147. Six years later, the same five-pack sold for $295, and by 2017 it was costing $530. Ninety percent of the global supply of the hormone was being made by Eli Lilly, Novo Nordisk, and Sanofi. Eli Lilly told Vice News that their net cost to make a box of insulin pens was $122. That included "manufacturing, labor, research and development, regulatory fees, promotional expenses, insulin donations and profits." In May 2019, Lilly introduced the generic version of its insulin (same drug, same packaging, and the same manufacturer), and they cut the price in half. The head of manufacturers who appeared before Congress, argued that their companies were giving discounts to pharmacy benefit managers, middlemen and women who negotiate the price of drugs with manufacturers for private payers, government programs, and large employers. They claimed that the amount a person paid for their products would be lower if the discounts went to the patient.

For Americans who live near one of the nation's boundaries, it's possible to buy cheaper insulin by walking or driving over the border. There are special lanes in Tijuana for U.S. citizens to cross to Mexico and buy medications. A million Americans use them each year. The FDA told Vice News that it is "illegal" for individuals to import drugs into the U.S. for personal use. As a practical matter, customs allows Americans to bring in 90 days' worth of medications at a time.

In 2017, Bernie Sanders, a senator from Vermont, introduced a bill that would allow the importation of prescription drugs from Canadian pharmacies, as long as they meet certain safety standards. Bernie's bill, of course, went nowhere. In 2021, Lilly added some people on Medicare D to their 2020 program. It offers people with and without commercial insurance to get a card that allows them to buy most Lilly insulin products for $35 per month. Pointing out that 7.5 million Americans use insulin and diabetes is the seventh-leading cause of death in the U.S., Lilly's ad asserted "no one should have to struggle to pay for their medicine."

Insulin has become a relatively expensive commodity and, in my mind, the question is not why the drug costs more in the U.S., but rather why it costs so much less in Canada. Why do Americans treat needed medications as merely another commodity, and believe that in this country it's OK to make large profits from some of our poorest and neediest citizens?

CHAPTER FIVE

Transfusions

"If you prick us, do we not bleed?"
—SHAKESPEARE.

Before the 20th century, doctors rarely considered transfusing people who bled profusely. Blood solidifies when it is outside the body, and when fresh blood from one person's veins was infused into the veins of another, it often caused fever, kidney failure, and death.

In a test tube, blood separates into two components. Red stuff, cells, occupy the bottom half of the tube, and clear fluid, serum, fills the top half. In the early 1900s, a Viennese immunologist named Karl Landsteiner mixed his red cells with the serum of others and noticed that some combinations clumped and others didn't. He wrote a paper and in a footnote suggested there probably were several "types" of blood antigens. Then he and his students spent eight years working out the details. By 1909, they knew there were two populations of antigens on the surface of red cells and two populations of antibodies in the serum. They labeled them A and B and discovered that:

- When red cells have B antigens on their surface, their serum contains antibodies that destroy A red cells.

- When the cells have A antigens on their surface, their serum contains antibodies that break up B red cells.

- When the exterior of red cells are not covered with A or B antigens, their serum has antibodies that break up both A and B red blood cells.

- When red cells have both A and B antigens on their surface, their serum does not contain A or B antibodies.

During the next few years Landsteiner, a significant Austrian researcher, "discovered how to infect monkeys with the syphilis bacterium and he helped prove polio was an infectious disease," but his blood group findings didn't seem to have a practical application.

Karl was six when his father's life ended, and he was forty when his mother died. He was so devoted to her that he "had her death mask taken" and hung it on his wall for the rest of his life. He married eight years after her demise.

In 1914, the rulers of the great European empires declared war, and 9 million young soldiers killed one another. When the war started, doctors occasionally transferred blood from one person to another with syringes or rubber tubing.

In 1917, the U.S. entered the Great War and Oswald Robertson, a doctor from Fresno, California, was sent to the front. He had learned about Landsteiner's work on blood types and knew that someone had recently discovered that citrate keeps blood from clotting. While caring for wounded troops, he collected blood in glass bottles, added sodium citrate to keep it from solidifying, and stored the liquid in ice. During the war, he transfused hundreds of wounded soldiers. Near the end of the conflict, he was teaching doctors from other units how it was done.

After the war, Robertson became a pneumonia researcher in New York. In 1923, he helped develop a school of medicine in China. While there, he developed a bad case of typhus, and when he recovered he returned to the U.S. He continued to be a researcher for most of his life, but he apparently was never again interested in blood or transfusions. He spent the last years of his life working in a laboratory in California's Santa Cruz Mountains where he studied the death of Pacific salmon.

During the two decades that followed the Great War, there were no blood banks and probably not many transfusions.

When the war ended Karl Landsteiner was living in Vienna. It had been the capital of the vast Austro-Hungarian Empire and was on the losing side of the First World War. The imperial lands were carved into many of the nations of modern-day Europe and there were shortages. That winter, Landsteiner's laboratory wasn't heated and a group of poor people "cut down the trees around his house for firewood and tore away his fences." Feeling personally threatened, Landsteiner moved to Holland with his wife and children. During the next three years, he performed experiments and was assisted by a manservant and a nun who was "very devout and frequently quit the lab for prayers or to serve as an organist in the chapel." A year or so later, Landsteiner accepted a position at the Rockefeller Institute and moved his family to New York City. On the ship that crossed the Atlantic, he told another passenger how much he loved living in the "little cottage with a rose garden" in the seaside town of Scheveningen, Holland. When he got to New York, he was surrounded by a new and different environment. He lived "on the floor above a butcher shop on a street with trolley cars." Avoiding social activities, he spent his days in the lab, and read and thought at night "until the late hour." In 1930, he received the Nobel Prize. In 1937, Alexander Wiener added another red cell surface antigen, the RH (Rhesus factor), to his equation.

The first civilian blood bank was set up by the Russians in 1932. Five years later doctors at Chicago's Cook County Hospital were the first Americans to "save and store" donated blood. In 1941, five months before the U.S. entered the Second World War, San Francisco's Irwin Memorial blood bank was established.

When blood, plus a chemical that prevents clotting, is put in a test tube, the cells settle to the bottom and the plasma floats to the top. For blood loss or significant anemia, packed red cells or erythrocytes are transfused. Each tiny disc lives for 120 days. In transfused blood, half the red cells are new and half old, so the average red cell in a unit of blood lasts 60 days.

In a blood-filled test tube, just above the red cells there's a thin layer of white cells and an even thinner stratum of platelets. White cells can live for up to two and a half weeks and platelets last an average of eight to nine days. White cells are an important contributor to our defense against infection, but in transfused blood they can cause adverse reactions.

Platelets are particles that plug holes and help stop bleeding. Some chemotherapy drugs can significantly suppress their blood levels for a number of days. When their absence creates a risk of bleeding that is high enough, platelets are gathered from others and given to people who lack them. Blood banks have machines that collect blood from volunteers, separate and collect the platelets, then push the platelet-poor blood back into the donor.

A year and a half before the U.S. became combatants in the Second World War, London was being bombarded by Nazi planes and the New York City Blood Donor Bureau started to stockpile "Blood for Britain." They accumulated thousands of units of blood, separated the cells from the plasma, and under sterile conditions shipped the fluid across the Atlantic. It was a huge undertaking, and the man in charge had previously only organized a group of people once. As a young man, he coordinated the paper routes of 10 childhood friends who were delivering two thousand newspapers a day.

When he was still a trainee, Charles Drew, the doctor who coordinated the effort, studied the preservation of blood products. He wrote a doctoral thesis titled "Banked Blood," and he knew how to produce plasma that had a two-month shelf life. Gathering, transporting, and processing thousands of units of blood was a complex undertaking, but Drew pulled it off and was able to send close to fifteen thousand pints of the precious fluid to the Brits. A black man, Drew was born in Washington, D.C., and was an outstanding high school athlete. In 1926, he was Amherst University's most valuable football player, and he went to medical school at McGill University in Canada. He graduated in 1933, and in 1941 he became the director of the first U.S. Red Cross blood bank. It was a big honor, but he didn't stay very long. He resigned because the organization labeled each unit of blood with the donor's race and didn't give blood from a black donor to a white patient. Drew is credited with saying, "No official department of the Federal Government should willfully humiliate its citizens; there is no scientific basis for the practice; and people need the blood." Drew returned to Howard University and became the chief surgeon at Freedmen's Hospital.

By the time I entered med school (1958), blood drives were coming to my campus annually, and I was a donor twice. People who were hemorrhaging or very anemic often needed transfusions. The Red Cross proudly boasted that it saved the lives of many wounded servicemen and women. By 1962, the year I graduated from medical school, there were already 4,400 hospital blood banks and 178 Red Cross and community facilities. I never knew what medicine was like before transferable blood was readily available.

In 1997, several San Francisco Bay Area blood banks merged and called themselves Blood Centers of the Pacific. The nonprofit corporations collected huge amounts of blood (two hundred thousand units a year) from willing voluntary donors. They then checked each unit for blood type and for disease, fractionated the fluid into its various components, and sold it to more than sixty hospitals. Their annual budget exceeded $40 million.

The blood supply is relatively safe in part because of the outrage of an angry man. In the '70s, a California legislator named Paul Gann capped our property taxes. That made him famous. But the legislation that bears his name, the Gann Act, has nothing to do with taxes. It deals with transfusions. It seems that around 1982 Gann had heart surgery and was transfused. Five years later, he discovered he had HIV. The blood he had received came from someone who was infected with the AIDS virus. Either the blood donor had not been adequately screened or the blood Gann received was not tested carefully enough. Gann was furious and apparently felt "there oughta be a law." So he wrote one.

Prior to elective surgery, California doctors must tell patients that they can store their own blood and have it available should they require a transfusion. Stockpiling blood prior to planned surgery can be tedious and costly. But it's intuitively better to get your own blood back than it is to receive that of another. It's also the law, so if the patient wanted it, we did it. The act also says people can refuse blood from the "bank" and, instead, get it from a donor they designate. The idea makes sense, but the blood from a friend or loved one is no longer safer than banked fluid. Before a unit of blood is given, it must be tested for the usual suspects, and it's logistically near impossible to collect, check, and process designated blood in an acute or urgent situation.

Before Gann's outrage, some blood bank executives argued that if they looked at blood too carefully, they would have to reject many donors and throw away too many units. Doctors wouldn't be able to treat the ill and people would die. After the Gann incident, blood banks (which were pretty good at questioning people about risk factors) got serious about screening blood for HIV, HTLV, hepatitis B, and hepatitis C. They also checked for a few other illnesses such as mosquito-borne West Nile virus, Zika, and cytomegalovirus. They screen for Chagas, a parasitic disease whose normal habitat is Central and parts of South America; and for babesia, a parasite found in New England that is transmitted by ticks.

We're apparently NOT yet testing the 11 million units of blood Americans use each year for dengue, a disease transmitted by mosquitoes that's common in Southeast Asia, or for chikungunya, a West Africa viral illness that was responsible (between 2014 and 2016) for the fever and joint pain of four thousand American travelers, most of whom had recently visited a Caribbean island. We also don't test for hepatitis E, the most common type of hepatitis in India and parts of Asia.

Before 1996, blood banks identified viral diseases by checking for the presence or absence of specific antibodies in the serum. When a virus invades a body, the immune system reacts and makes detectable antibodies. It was believed that blood that did not contain certain antibodies should not be infectious. To prove their blood was safe, blood banks participated in studies on people who were transfused with blood whose antibody levels had been tested. Approximately 2.3 million transfusions were given during the study period, and people were subsequently evaluated to see if they remained disease free. One in every 493,000 infused units caused HIV, hepatitis C was seen after one in one hundred thousand transfusions, and hepatitis B one in sixty-three thousand. Screening was good but imperfect. During the early weeks after a person is infected, the virus incubates and its numbers grow. It takes a while before measurable antibodies develop, so blood can be contagious when the antibody tests are negative.

Over time, PCR technology improved and we were able to directly detect and measure miniscule amounts of virus. (PCR is like a Xerox machine for DNA. It allows technicians to make millions of copies of the original, to turn a tiny amount of stuff into a wad large enough to analyze and learn what we are dealing with. During the COVID-19 epidemic, PCR technology is being used to help identify the virus in nasal passage secretions.)

In 1999, blood banks started using the technique to screen all 66 million units of blood that were transfused. Between 2006 and 2008, with PCR testing being used, the recipients of 3.5 million units of blood were checked to see if they had been infected with any of three common chronic viral diseases. One in 1.85 million units of blood that was free of "measurable" viral particles caused an HIV infection, one in 246,000 transmitted hepatitis C, and one in 410,000 gave the recipient hepatitis B. We're not perfect yet.

While blood is donated freely, screening the donor and acquiring, testing, and distributing the red stuff is expensive. A recent survey put the cost of a unit of transfused blood at $522 to $1,183. In most hospitals, much of the blood is used at the time of surgery. Hospitals vary in size and in the numbers and the types of operations performed. So it's not surprising that, in the same survey, acquired blood cost $1.6 million to $6 million per hospital annually.

It's not illegal to pay people who donate blood, but hospitals won't use it. Studies a few decades back showed that people who receive paid units have an increased risk of developing post transfusion disease. Blood banks do, however, pay donors for plasma, the straw-colored liquid portion of blood. The U.S. exported $19 billion worth of the yellow fluid in 2016. According to a leading ethicist, plasma is "a very valuable commodity" that "like oil has to be refined and moved and shipped." A recent book claims about half of the plasma in Europe comes from the U.S.

CHAPTER SIX

Viruses

In the late 1800s, years before we could visualize a virus, a Russian scientist proved there was a life form that was too small to be seen with a microscope. He gathered liquid from the infected leaves of tobacco plants and passed it through a porcelain filter. The filter's porous openings prevented bacteria and fine particles from getting to the other side, but the fluid that trickled through the holes contained something that infected tobacco leaves. A very small entity had slithered through. We could only infer the existence of a virus before Ernst Ruska, a German physicist, found a way to "observe" the life forms. Ruska's lifelong pursuit of a powerful magnifying device started when he was a child and his father, a passionate botanist and mineralogist, forbade his sons from touching his precious microscope. As an adult, Ruska learned that it's not possible to produce an image when an object is smaller than the wavelength of light and he knew electrons have an even shorter wavelength. In the early 1930s, while earning his Ph.D., Ruska learned how to employ magnetic coils as a lens for electron beams. Using a number of the coils he developed an electron microscope and was able to capture images of viruses.

Once we could visualize viruses, we learned they cause a large number of minor, serious, and occasionally chronic viral diseases. Vaccines were developed that protect us from many of the diseases they cause, but outbreaks of influenza, dengue, and HIV continue to plague mankind.

In the fall of 2020, the president of the United States was infected by COVID-19, a unique coronavirus that emerged from a bat. It was spreading through the world and killing 5 percent of the vulnerable elderly. President Trump had a cough, fever, and a low blood level of oxygen. After a few days of waiting it out, he was increasingly symptomatic and was helicoptered to Walter Reed hospital. While there (among other treatments), he received dexamethasone, a potent steroid and infusions of antibodies to the virus and solutions that contained the antiviral drug remdesivir. The use of drugs to suppress the level of a virus in a person's body isn't new. In the past, these medications have helped people who have an active inflammation with hepatitis B or HIV or who are prone to recurrences of herpes. Remdesivir is an antiviral drug that is administered by five infusions over five days. As I see it, giving infusions and lowering the level of virus in a person's body may buy a person time to develop protective antibodies. While they are waiting, it prevents lung, heart, brain, or vascular problems.

Remdesivir was not approved by the FDA for general use at the time, but the agency issued an emergency authorization. It allowed doctors to give the medication to people physicians believed were sick enough. On October 22, 2020, the drug was approved by the FDA for general use in the hospital.

The medication was developed by Gilead, a California pharmaceutical company that in the early years of the 21st century managed to acquire some of the world's most effective treatments for HIV, hepatitis B, and hepatitis C.

The corporation was created by Michael Riordan, a young man who spent the year after he graduated from college as a volunteer in Asia. While there, Riordan contracted dengue fever. Dengue is a viral disease that is transmitted by mosquitoes. It is "a leading cause of serious illness and death in some Asian, African, and South American countries." Riordan was "flat on his back for three weeks." At the time, there was no drug treatment for most viral diseases, and that might have been part of what motivated Riordan to create a start-up that developed antiviral drugs.

Riordan had grown up in Wichita, Kansas, and was the son of a physician who was a "medical maverick," a man who "brought a unique perspective to the world of medicine." In college, Michael was a swimmer and one of the best students in his class at Washington University in St. Louis. When he graduated, Riordan turned down a Fulbright fellowship and accepted one created by Henry Luce, the man who was raised in China and who founded *Time* magazine.

After his year in the Far East, Riordan became an M.D. at Johns Hopkins, and an MBA at Harvard. When he was in his twenties, he found people who were willing to invest millions in his plan to produce antiviral medications.

In 1987, with $2 million in seed capital, six employees, and little research experience, Riordan convinced former secretary of defense Donald Rumsfeld and a few others to join Gilead's board. In 1990, he reeled in John Martin, the director of infective chemistry at Bristol Myers. Martin was highly thought of because he had helped create ganciclovir, an antiviral that controls herpes.

While still an employee of Bristol Myers, Martin learned about a special group of viral-fighting chemicals called "acyclic nucleotide phosphonates." The promising drugs had been developed in Prague by Antonin Holy, a "tenacious and meticulous chemist," and had been evaluated in Belgium by a researcher named Erik De Clercq. They were said to be effective destroyers of DNA and RNA viruses, and Bristol Myers was planning to license them.

Viruses are chains of normal nucleotides, and most antiviral drugs are chemically modified nucleosides. They work inside cells, mimic normal nucleotides, and block viral replication.

By the late 1980s, Holy had been sneaking promising chemicals out of Czechoslovakia for twenty years. He sent or gave them to De Clercq in Belgium, and De Clercq tested the substances and learned many seemed to be effective.

After a Prague uprising in 1968, a number of rules and regulations restricted trade between Czechoslovakia and the West. But Czech merchants were still able to export the hops that Belgians used to make beer or to import powdered Belgian milk. In 1981, when De Clercq visited Prague, Holy took his friend to dinner and stuffed his coat with vials of newly developed compounds. Somehow the chemicals got through.

Martin was an associate director of antiviral chemistry at Bristol Myers and decided to visit the Czech chemist his company planned to work with. He flew to Prague, met Holy, and as the two strolled through the city's narrow streets and cobblestone alleys, they got to know and like one another. Holy wasn't a complainer. He worked with homemade reagents and without laminar flow hoods to get rid of fumes. His lab was "as plain as a kitchen," but he managed, and he had no desire to leave the country of his birth. When Martin returned to the U.S., a CIA officer tried to get Martin to turn Holy into an "asset" and Martin declined.

In July 1989, Gorbachev allowed the nations of the Eastern Bloc to go their own way. The iron curtain fell and commerce and travel between the East and the West became relatively easy. In the U.S., Bristol Myers merged with Squibb, and the company's leadership and goals changed. In mid-1991, with the head of Squibb calling the shots, the newly formed mega company decided they weren't interested in Holy's chemicals. They didn't want to take the risks and make the investment necessary to perhaps develop another HIV medication. Realizing Squibb's error, Martin phoned Holy and convinced him to sign a licensing agreement with Gilead. In July 1991, Holy, Martin, and De Clercq met in a restaurant near the Eiffel Tower and signed a deal on a napkin. Holy's nucleotides have become the basis of tenofivir, one of the most potent medications (and moneymaking commodities) that are currently used to prevent and treat HIV.

It took fifteen years before Gilead was profitable. During that time, the company spent $2 billion and lost three-quarters of a billion.

In 2011, Gilead bought Pharmasset for $11 billion and gained control of the start-up's potent hepatitis C medication, sofosbuvir. When the drug was given with a less potent antiviral the combination quickly cured most people who were chronically infected with hepatitis C. It was an amazing breakthrough and was great news for the 2.7 to 3.9 million Americans and the 71 million worldwide who carry the bug. Some people who were chronically infected with the virus developed cirrhosis or liver cancer, and hepatitis C was responsible for the deaths of four hundred thousand people each year. Before it was identified in the '70s by a Chiron team led by Michael Houghton, we didn't know what caused most cases of transfusion hepatitis. Hepatitis C turned out to be one of several viruses that inflames the liver, turns a person's skin and eyeballs yellow, and drains their energy. It becomes chronic in 70 to 85 percent of those who acquire the disease when they are adults. A third of the chronically infected aren't visibly ill but have a disease that smolders. These individuals usually develop cirrhosis and die within twenty years. A second infected group never has significant problems. And in a third, the virus isn't harmful for decades but at some point, for some reason, the virus starts to slowly and progressively inflame the infected person's liver.

We know of four other viruses that affect the liver, and each behaves a little differently. Hepatitis A doesn't become chronic; hepatitis E can become a persistent problem in people who are immune-suppressed; and hepatitis B usually causes a self-limited illness in newly affected adults, but becomes a lifelong problem for the infants who acquire the disease from their infected mother.

In the decades after it was identified, hepatitis C was treated and often cured with interferon. The treatment was a year-long ordeal and consisted of weekly injections that cause fever and exhaustion. The bad effects usually lasted a few days and subsided before it was time to get the next shot. I recall telling a young man who was being treated: "If you get interferon on a Friday, you will be sick on Saturday and Sunday, but will probably be well enough to go to work Monday. Would you like to be treated that way?" "And ruin my weekend?" He shook his head. "I don't think so."

My colleagues and I treated hundreds of people using this regimen. The process was emotionally trying, but the people who desperately wanted to be cured endured the weekly draining days. In a significant minority, the treatment didn't eradicate their disease, and for many that was heartbreaking.

Raymond Schinazi, the doctor who oversaw the development of the curative drug, was a Jew, born in Alexandria, Egypt. When Israel and Egypt fought a war in 1956, Egyptian Jews became personae non grata. His family moved to Italy, and he later studied at Bath in the U.K. "As a student he lived on one hundred English pounds a month, worked as a parking attendant to help pay his way, and didn't have real money in his pocket until he won three thousand pounds in the Spanish lottery." After receiving his British degree, he did his post-doc work in Yale and spent three years making "chemicals similar to nucleosides." Described as a "bear-sized man who speaks bluntly, negotiates fiercely, and favors splashy, multicolored shirts," Dr. Schinazi enjoyed the science but not the weather at Yale. "We had two really cold winters in a row in New Haven, with tons and tons of snow."

When the world learned HIV was caused by a virus, Schinazi was a professor of pediatrics at the VA hospital in Atlanta. As he explained (in interviews), he "couldn't just sit around and do nothing." He wanted to attack the virus with nucleoside analogues and the VA resisted. Eventually, Schinazi helped develop two of the more significant anti-HIV drugs and made sure Emory University profited from the sale of the medications.

Encouraged by his success, Schinazi wanted to try to develop an anti-hepatitis C drug. The National Institutes of Health (NIH), allegedly, turned down his application for the project. Schinazi and a partner got venture funding and founded Pharmasset. One of their company chemists, Michael Sofia, developed sofosbuvir, a drug that, with a little help, almost always cured hepatitis C in twelve weeks.

Pharmasset planned to sell the drug at a total cost per treatment of around $30,000, but in 2011 Gilead bought Pharmasset for $11 billion. Schinazi received $440 million and went on to do further research. Gilead now had to charge a lot for the medication. The original list U.S. price for the company's two drug combination, Harvoni, was $94,000.

During an interview with journalist Jon Cohen, Shinazi was asked about "sofosbuvir's price tag of $1000 per pill." Shinazi pointed out that Gilead decides on the price, and he called it "obscene" but not unreasonable. "Is it fair to pay $3 for a bottle of water when you're thirsty? This is something that cures you from a disease."

Once Gilead owned the antivirals, it spent $60 to $80 million on TV ads. Actors faced the camera and said they were "ready" to be cured. The company needed to sell a lot before they faced competition, before the medication became one of several low-cost commodities. In 2014, Gilead's drug combination, Harvoni (ledipasvir/sofosbuvir), created revenues of more than $10 billion.

In 2002, a coronavirus emerged from its animal host in China and set off an epidemic of "severe acute respiratory" infections (SARS). Eight thousand people were ill, more than eight hundred died, and the outbreak lasted for a year and a half. People were not usually contagious before they got sick. Isolating people with symptoms, quarantining contacts, and restricting travel were thought to be the main reasons the virus stopped spreading.

Less than a decade later, a different bat coronavirus attacked camels on the Arabian Peninsula. Then it spread to man. Called MERS, Middle East Respiratory Syndrome, it caused a series of respiratory infections that killed eight hundred people. It was highly lethal, but the vast majority of the people who got sick had "unprotected contact with sick camels. Some were living with or caring for people who had the disease and others were infected in the Middle East and then went abroad." MERS never became a major worldwide epidemic.

At some point, the U.S. government asked Gilead to search its library of a thousand small-molecule nucleoside analogues. They wanted to know if Gilead had a medication that could potentially be used to treat infections caused by RNA coronaviruses. Drug number 5,734, remdesivir, turned out to be broadly active.

In late 2019, a novel, sometimes lethal, coronavirus started in China and spread throughout the world. This one is usually contagious before it makes people sick and it doesn't look like it's going to burn itself out.

In May 2020, remdesivir was given to half the people who had a severe case of COVID-19. It was administered on average nine days after their infection started and the people who were given the medication were quite sick. They were treated in a hospital and those who got remdesivir recovered four days sooner than the people who received a placebo. In October 2020, the drug was infused into the body of an ill U.S. president, and on October 22 it was approved for general use in ill people. Uncontrolled studies in Europe suggest the drug doesn't help very sick hospitalized patients.

According to NPR (June 29, 2020), a typical five days of infused remdesivir has been priced at $3,120, which, in today's world of very high drug prices, actually sounds quite reasonable.

CHAPTER SEVEN

Vaccines

During the 1800s, Pasteur, Jenner, and others used live viruses to produce vaccines. Their creations protected people and animals from smallpox, rabies, animal anthrax, chicken cholera, and probably a few other diseases.

During the First World War, eight of every one thousand injured British soldiers died or were sickened by a poison that was made by *Clostridium tetani*. Their condition was called tetanus, or lockjaw. The germ that made the toxin often exists in dirt. It thrives in the absence of oxygen, and gets into wounds. Eight days after a soldier was injured, the trapped bacteria had usually generated enough toxin to cause jaws to tighten, muscles to become rigid, and breathing and swallowing to become difficult.

In the late 1800s, a Scottish toxicologist created a serum that protected animals from the venom of a cobra. Later a Japanese physician developed an antiserum that prevented "lockjaw." The Japanese investigator, Kitasato Shibasaburo, was born in a mountainous village on the southern Japanese island of Kyushu. His government sent him to Berlin and he worked in Koch's lab. While there, he became the first researcher to grow a pure culture of the tetanus bacillus. In 1890, he and another investigator "injected sub-lethal doses of tetanus toxin into rabbits" and produced an antitoxin. It allowed doctors to block the effect of the venom, but it didn't prevent the condition.

In 1924, a French veterinarian named Gaston Ramon inactivated the deadly tetanus poison with formaldehyde and added an "adjuvant," a chemical that boosts the immune response. When his creation was injected into a person, the recipient developed antibodies that blocked the tetanus toxin. Ramon also helped develop the vaccine that opposed the diphtheria toxin. It prevented the thick throat and nose membrane that, in severe cases, obstructed airways and made breathing difficult. Each of Gaston's creations saved countless lives and he was nominated for the Nobel Prize 155 times, but he never won.

By the time the U.S. entered the Second World War, the U.S. Army was routinely using the vaccine for tetanus. Though there were over 2.5 million wounded soldiers during the conflict, only twelve developed lockjaw. Four of those individuals probably didn't get all three shots.

In the early 20th century, researchers learned how to use viruses that were killed with formaldehyde to create immunity to various diseases. More than forty of these types of vaccines were developed by Maurice Hilleman. The son of a farmer, Maurice was born on a hot summer day on a Montana farm near the banks of the Yellowstone and Tongue Rivers. His mother had a severe complication of pregnancy called eclampsia. Before she died, she gave her two-day-old newborn to the childless couple down the road. Maurice claims he was put to work as soon as he was able to tell a weed from a seed. "Everyone in Montana had to earn their keep." On the farm, they sold what they grew. He picked berries and watered the animals. As a teenager, he worked at J. C. Penney in the nearby town and "helped cowpokes pick out chenille bathrobes for their girlfriends." He decided he didn't want to go to the local college because he "didn't want to be strapped down by the church dogma." He won a scholarship to Montana State University and graduated first in his class. That got him into the Ph.D. program in microbiology at the University of Chicago. "Despite receiving a scholarship, money was always in short supply. Hilleman lived in a squalid apartment and survived on a single meal each day. At six feet, one inch tall, he weighed less than 140 pounds."

As he explained, the Chicago University professors didn't want to be bothered unless "you discover something." While in school, he identified an organism that was sexually transmitted. It infected 3 million

Americans a year and scarred many fallopian tubes. It was a bacterium called chlamydia. It only grew inside the cells of the host and it was easily destroyed by antibiotics. Hilleman's professors tried to make him stay in academia, but he "came from a farm and wanted to do something—to make something."

In 1944, as a new employee of Squibb, Hilleman developed a vaccine for Japanese encephalitis virus. The virus infects pigs in parts of Southeast Asia and is transmitted by mosquitoes. In humans, the infection is usually mild but it sometimes causes fever, seizures, and serious long-term movement problems. In 1944, the war in the Pacific was heating up and the U.S. Army wanted a vaccine that would protect soldiers. Hilleman offered to make the vaccine for $3 a dose and he won the contract. Squibb gave him a horse barn and an engineer. An old farm boy, he bulldozed out the manure, painted the floor, and went to work. From his U. of Chicago days, Hilleman knew the encephalitis virus would grow in the brain of a mouse. As he later explained to fellow "vaccinologist" Burt Dorman, making a vaccine is like getting the old tractor working. You fiddle with it. If that doesn't work, you fiddle with it some more. Hilleman figured out how to create the vaccine and assembled a crew of thirty women. They injected the virus into the skulls of mice and waited for it to grow. After a few days, they killed the rodents with ether and "harvested 30,000 brains a day." The organs were homogenized with a blender, washed, centrifuged, and washed again and again. The process was repeated until the solution contained pure viruses. Then the infectious agents were inactivated with formaldehyde. Three months later, Squibb had enough vaccine to immunize two hundred thousand troops.

After the Second World War, the U.S. military and the World Health Organization were always on the lookout for the next influenza pandemic. In 1957, Hilleman was working at Walter Reed military hospital and he read about a flu outbreak in Hong Kong. It infected 10 percent of the population—250,000 people—so it was worrisome. He contacted an army physician in Japan and the doctor got an infected navy serviceman to gargle and spit into a cup. A month later, the specimen arrived in D.C. and Hilleman put the spit into a fertilized egg. Then he watched the flu virus grow and developed a way to test for the presence of antibodies to the germ. He checked the serum of hundreds of people and found that very few living humans had been exposed to anything like this virus. The only people who had antibodies were 70 and 80 year olds who had survived the flu epidemic of 1889–90.

Hilleman figured that a virus would arrive in the fall and create a devastating epidemic, and he "sent out a press release and warned the world" but no one seemed to believe him. "Joe Bell," a prominent U.S. Public Health Service epidemiologist known for his attention to detail, commented, "What pandemic? What Influenza?" "How," Hilleman wondered, "could people live in this world and be so stupid." Eventually, Hilleman went to a restaurant where the head of the U.S. influenza commission was eating. He interrupted the man's meal, showed him the information, and told him that ignoring his findings was "a big mistake." The commissioner checked the data and agreed. It was a pandemic virus.

Hilleman contacted six companies that produced flu vaccine. He cut the red tape, sidestepped a few rules, and told chicken growers NOT to kill roosters. As a farmer, he knew roosters were usually killed late in the hatching season and companies would need a lot of fertilized eggs. By late fall, 40 million doses of vaccine were available and many people were immunized. The disease arrived in the U.S. that September in the body of a girl who attended a church conference in Grinnell, Iowa, and in the lungs of Boy Scouts from Hawaii who attended a jamboree at Valley Forge, Pennsylvania. The 1957 flu killed seventy thousand Americans and 4 million worldwide, but thousands of lives were saved.

In the 1950s, Hilleman moved to the Merck Company and started a revolution of sorts. Over the years, his group turned out products that prevented a staggering forty animal and human diseases. (Paul Offit, M.D., a professor of pediatrics detailed Hillman's story in his excellent book *Vaccinated*.)

The vaccines for measles, mumps, and chicken pox used viruses that were grown in non-human cells, mutated, and induced immunity but were not infectious. They became available in the late '60s and early '70s. In 1957, the year before I graduated from med school, the people I interviewed all remembered their childhood illnesses, staying home for a week, with fevers, rash, and a swollen jaw.

In the 1950s, Jonas Salk, using inactivated viruses, developed a vaccine for polio. The studious son of immigrants, young Jonas was "a perfectionist" and a good student. After medical school he became a research bacteriologist, and in 1941 he began assisting a forty-one–year-old influenza researcher named Thomas Francis. The son of a steelworker, Francis had spent the previous decade studying the influenza virus. The nation remembered how the pandemic of 1918–19 had killed more soldiers than their wartime enemy, and the government didn't want a repeat. The U.S. Army asked Francis to develop an influenza vaccine. As one of his helpers, Salk learned how to create inactivated vaccines and how to prove they worked.

In 1947, at age thirty-three, Salk was given a laboratory at the University of Pittsburgh and was funded by the March of Dimes, a charity that was created and promoted by Franklin Roosevelt. Roosevelt was the president who led the country during the depression of the 1930s and for most of the Second World War. When he was thirty-nine years old, he developed paralytic polio and he was unable to walk without help. His disability was public knowledge and the fear of the virus was heightened by a 1952 polio outbreak that caused 3,145 American deaths.

Salk's vaccine was created by chemically inactivating live virus. In the early 1950s, it was given to 2 million schoolchildren as part of a placebo-controlled trial. In April 1955, the nation learned that it was safe and effective and Salk became an instant icon. His celebrity status troubled his medical colleagues and affected his marriage of twenty-eight years. He ended up divorcing his first wife and married Francoise Gilot, a woman who had been the "longtime lover of Pablo Picasso."

When famed journalist Edward R. Murrow asked Salk, "who owns the patent on this vaccine," Salk replied, "the people." Then he paused for a moment: "Could you patent the sun?" Some claim the University of Pittsburgh lawyers had already concluded the product wasn't patentable.

Salk founded a research institute in San Diego. He was interviewed, and in a book was psychologically picked apart. His interviewer wrote that the man "movie stars came to visit and babies were named after" was "conceited but vulnerable—mild-mannered, but often arrogant and combative."

The other developer of a polio vaccine was born in Poland and was fifteen when he and his parents came to the United States. Albert Bruce Sabin eventually went to medical school, graduating in 1931. While studying polio at the Children's Hospital Research Foundation in Cincinnati, Ohio, Sabin discovered that polio viruses lived in the small intestines before they attacked nerves. During the Second World War, Sabin served in the army and developed at least one vaccine. When the conflict ended, he returned to Ohio and resumed his research. By growing polio virus in monkey kidney and testicular cells he created a polio mutant that flourished in the intestines, didn't cause disease, and prevented infections with the wild virus. He then cultivated and tested a few strains of his virus and gave doses to his family, and to prisoners from a nearby penitentiary. The viruses seemed safe and they became the basis of an oral vaccine. At the time, the U.S. wasn't interested. The country was committed to Salk and wouldn't allow Sabin to test, produce, or administer his creation.

In 1955, the Russians founded a Polio Research Institute in Moscow. In 1956, its head virologist, Mikhail Chumakov, visited Salk in Pittsburgh and Sabin in Cincinnati. Sabin spoke a little Russian, Chumakov a little English, and they became friends. The following year, Sabin spent a month giving lectures in Russia, meeting with researchers, and promoting his vaccine. When he returned home, the Salk vaccine was being widely used and an American researcher advised Sabin to toss his vaccine "in the sewer." In 1958, Chumakov tested a vaccine made with Sabin's seed virus on twenty thousand Russian children. It was safe, easy to administer, and effective. Chumakov got permission from the public health chief of the Politburo and distributed the oral vaccine to more than 15 million Soviets. He later treated 23 million children in East Germany, Czechoslovakia, Hungary, Romania, and Bulgaria. An expert sent by the WHO agreed the vaccine seemed to be working but "definitive results would take time." Sabin refused to patent the oral vaccine, and it is currently widely used in the poorer countries of the world.

On April 26, 1955, shortly after it was licensed, the nation learned the Salk vaccine was making some kids sick and was causing paralysis and a few deaths. Half of the 760,000 doses made by Cutter in Berkeley, California, were recalled. In the end, "forty thousand kids who received the vaccine developed polio, two hundred had at least partial paralysis and ten died." The formaldehyde in the vats had failed to completely inactivate the virus.

The vaccine was pulled from the market. Six years later, the Sabin vaccine became available in the U.S. and was widely used. During the following decades, the oral virus would occasionally mutate in a person's intestine and cause polio and paralysis.

In 1987, a new, potent injectable polio vaccine was introduced. It was safer than the original Salk product and it became the only vaccine used in the U.S. and Europe. In the rest of the world, "more than 10 billion doses of oral vaccine were given to 3 billion kids during the last twenty years." The World Health Organization thinks they have prevented 13 million cases of polio, but the Sabin vaccine did cause a few cases of polio.

During the last half of the 20th century, states started requiring kids who attended public schools to be immunized. Most families complied and the rest were usually shielded by "herd" immunity. If all or most kids in a community are protected, those who have not been immunized are unlikely to be exposed to the disease. These shots sometimes caused fever and joint aches, and one in a million kids developed encephalitis, a serious brain inflammation. The complications led to lawsuits and pharma wanted to stop producing vaccines. In response, Congress passed the 1986 National Childhood Vaccine Injury Act, and the government started compensating the families of people who were harmed.

When the Cutter labs failed to effectively kill the polio virus, many became wary of killed virus vaccines. Then Baruch Blumberg showed that antibodies to the outer shell of a virus were sometimes all we need to protect people from a disease and we started thinking differently.

Blumberg was the son of a lawyer who struggled during the Great Depression. He was sixteen when the Japanese bombed Pearl Harbor and he joined the navy and became a midshipman. When the war ended, he used the G.I. Bill to pay for his schooling and planned to study math. But his father, who usually didn't say much, said, "I think you should go to medical school," and he said OK. He met his future wife, Jean, when he was a medical intern at Bellevue hospital. She worked in the hospital research lab and their dates were dinners in the hospital cafeteria and strolls through the streets of New York. "A child of the depression,"

Blumberg was paid $25 a week and managed to save most of the money. He became a researcher, and in the days before we knew how to access DNA, he tried to classify genes by the presence of certain unique proteins in a person's blood. (Many genes direct cells to make a specific protein.)

One of the oddball proteins he discovered clearly couldn't be genetic in origin. He found it in an Australian Aborigine, and it was also present in the blood of a New Yorker. The American had hemophilia and had been transfused many times. There was no way the two could have a unique genetic connection. Over the years, Blumberg studied the protein and learned it was part of the shell of the hepatitis B virus. When the virus replicates it is overproduced, and it spills into the blood.

When the protein was present in a person's blood, it indicated an active infection. When it was absent and the blood contained an antibody to hepatitis B, it meant the individual had been infected but was now immune.

Maurice Hilleman used the surface antigen and developed a vaccine. He was unable to completely eradicate the virus from the blood, so he treated the preparation with formaldehyde, pepsin, and urea. That should have killed any virus that was present. To convince the public it was safe, he asked Merck mid-level executives to be the first to be injected. As Hilleman later quipped, "We found the most worthless people we could find in the world who would be least likely to sue."

No one volunteered so he gathered the execs and explained that "No" was not an option. Fearing the vaccine might have been contaminated with HIV, some of the executives who were injected later admitted they took the shot but were "scared to death."

When he later learned the vaccine production was being accelerated, Hilleman met with the people making it. His killing process was effective, but it was slow. Hilleman assumed steps were being skipped and no one knew if it affected the vaccine's safety. He told the production crew he suspected someone was "changing the fucking process so he can get more yield and earn a bonus. Meatheads are everywhere."

The world needed a safe and effective hepatitis B vaccine. The virus was present, though usually inactive, in the livers of over 200 million people in the world. Approximately 1.2 million were Americans. But people feared the Hilleman vaccine and didn't take it.

In the process of making his vaccine Hillman had demonstrated that it was probably possible to use the outer shell of a virus to make a vaccine and Bill Rutter took notice. In the late 1960s, Rutter decided to try to use genetic engineering to make the membrane that surrounds the virus. If he was successful, scientists could use the protein to develop a vaccine that people wouldn't be afraid to use.

Rutter was the chief of biochemistry at the University of California in San Francisco. He was the son of a British Quaker and was born and raised in Malad, a small town in southern Idaho. His grandfather had been a British Army officer in India and had told young William about the poverty and exotic parasitic diseases he had witnessed. After hearing the stories, young William thought he would someday go to the school of tropical medicine in Calcutta.

During the Second World War, Rutter lied about his age, and when he was 16 he joined the navy. When the war ended, he went to Harvard and was drawn to science. He was accepted to Harvard medical school, but after he attended a few medical school classes with a cousin, he realized he wanted to be a scientist. He earned a Ph.D. at the University of Illinois and spent a decade as a researcher and professor at the University of Illinois, Stanford, and the University of Washington.

In 1968, after refusing the job offer three times, he became the chief of the biochemistry department at the University of California in San Francisco. The unit had been leaderless for six years and (he once quipped) "every good scientist in the United States had probably been asked to take that job and turned it down." He claimed he accepted the post because twenty faculty positions were open, and that was a "bonanza for recruiting." In academia, investigators don't always cooperate and share. Rutter knew science was competitive. "Everyone is trying to beat you and will use every trick in the book. You try to cover your bets in many different ways." Rutter gathered top-notch researchers and got them to work collaboratively. In his department, associates shared their knowledge.

Herbert Boyer was a researcher at UCSF when Rutter was chief. Boyer learned how to insert genes into plasmids and get bacteria to make a desired protein. He shared his knowledge with Rutter before he left the university and helped create Genentech.

In the 1970s, researchers learned how to determine how the nucleotides lined up to form the blueprints for genes and viruses. When certain
genes—explicit lengths of DNA—were placed into plasmids and planted
into a bacterium, they forced the cell to produce a specific protein.

The idea of using implanted genes to make proteins frightened some
politicians. "Some portion of scientists was genuinely concerned. Others
enjoyed the debate and the public controversy."

There was a Senate hearing on the topic of genetic engineering and
Rutter attended. Margaret Mead arrived wearing a huge long robe and
carrying a shepherd's staff. "Adlai Stevenson, a Senator and a lawyer who
would later run for the presidency of the U.S., ran the proceeding. He
introduced Margaret Mead as a world-renowned scientist who could
give guidance on these issues." Mead was an anthropologist who became
famous after she spent nine months in Samoa. She learned that "adoles-
cence on the island was not a stressful time for girls because their cultural
patterns were different than those in the U.S."

At the hearing she stood and repeatedly said something like, "You're
going to hear today from these scientists that this (genetic engineering) is not dangerous. I'm here to tell you it is dangerous." After each
repetition of her statement, she pounded the floor with her staff for
emphasis. "Boom! Boom!" To Rutter, "a social anthropologist with her
shepherd's staff giving advice on molecular, microbiological, and phys-
iological science" was troubling. Observing the theatrics and attitudes,
Rutter realized that if he wanted to make the hepatitis B vaccine his
way, he would have to give up his job as chief of biochemistry and go
private. In 1981, Rutter acquired venture funding, hired great researchers, and formed Chiron. "It was not an issue of damn the torpedoes,
full speed ahead." The business of the company initially was research,
pure and simple—understanding the potential of a set of technologies.
The major pharmaceutical companies didn't want to become polluted
by something that was "controversial" like genetic engineering, and they
stayed on the sidelines.

Merck gave Rutter's group the hepatitis B DNA and one of Chiron's
founders, Pablo Valenzuela, isolated the section of the viral genome that
coded (was the blueprint) for the outer membrane of the hepatitis B virus.

The Chiron team inserted that DNA into a plasmid and put the plasmid into yeast. The organism made significant amounts of Hepatitis B surface antigen. When the particles were injected into people, their bodies produced antibodies that attacked the outer membrane of the virus, and the vaccine prevented hepatitis B. It was the first effective human vaccine made of viral sub-units but it was not the last.

By the 1990s, pharmaceutical companies had largely stopped using formaldehyde-inactivated viruses, and the immunizations they made were "technologically more sophisticated and substantially more expensive. Takeda in Japan used envelope proteins to make a 9-valent human papilloma virus vaccine. Papilloma is a viral infection that dramatically increases the future risk of cancer of the cervix. The vaccine is given to girls before they become sexually active. (See section on childbirth.)

For decades, insurance policies covered the costs of none, some, or all of the charges for new vaccines. Then the Affordable Care Act was passed, and on September 23, 2010, the president made it official. "Preventive services that have strong scientific evidence of their health benefits must be covered and plans can no longer charge a patient a copayment, co-insurance or deductible for these services when they are delivered by a network provider."

An advisory committee of the Centers for Disease Control published a list of twenty vaccine-specific recommendations for conditions ranging from rabies to smallpox.

In 2020, a novel, sometimes lethal, coronavirus kicked off an epidemic, and by June 2020, according to the website MarketWatch, nine pharmaceutical companies were racing to develop a vaccine that stimulates the body to develop antibodies to part of the viral shell. Millions of dollars were being advanced by the government, and investors, hoping to make a lot of money, were buying the stock of vaccine makers.

Shortly after the virus started ravaging China, researchers determined the sequence of nucleotides that led to the formation of the infectious agent. A section of the RNA was the blueprint for the spike proteins that speckle the wall of the coronavirus. The spike connects the virus to the cell it is going to invade. COVID-19 is an RNA virus. Two companies transformed a slice of viral RNA into a piece of messenger RNA and "packaged" the new RNA in tiny balls of fat called lipid nanoparticles. Other companies made DNA copies of the spike RNA and "added the 'gene' to an adenovirus that can enter cells but can't replicate inside them or cause illness."

When the viral particles, or adenovirus, is injected into a person, it enters cells and forces them to make spike proteins. The body's immune system recognizes the new proteins and realizes they are foreign antigens. Lymphocytes make antibodies. Some of the vaccines "activate helper T cells." So far it looks like the vaccines will prevent or modify most COVID infections, but the virus is slowly mutating. Scientists worry that if the virus changes enough, the antibodies the vaccines are creating won't recognize some of the new viral strains.

Researchers have been unable to make a vaccine that prevents HIV. Within two to three days after a person is infected with HIV a few of the RNA viruses have already entered a T lymphocyte. They have created DNA copies of themselves and the DNA has become a gene. The enzyme the viruses use to make DNA copies of the RNA (reverse transcriptase) is error prone. When viruses make DNA copies they make mistakes. At least one DNA base pair is incorrect and the virus that's created is a mutant. When infected people make antibodies, their immune systems can't keep up with the viral alterations, and most HIV antibodies are not protective. (A rare person eventually makes broadly neutralizing antibodies that are active against a wide variety of different viral strains. They are able to control the HIV infection.) The drugs that fight HIV don't cure the disease because they are unable to kill the occasional infected cell that is inactive but that wakes up when people stop taking antiviral drugs.

CHAPTER EIGHT

Surgery Becomes a Learned Craft

I n 1904, William Halsted started the nation's first surgical residency program. He was a doctor who was born into wealth and privilege. His family had a home in New York City, a second home on the Hudson River, and he was educated by a private tutor. He graduated from Yale without ever having checked a book out of the library, and he was the senior captain of the football team. He had no problem getting into medical school or taking off a year mid education to sail and fish. When other medical school graduates were surgical apprentices or starting a practice, he was an intern at Bellevue hospital and later a house physician at New York Hospital. Then he spent two years in Germany, Austria, and Switzerland learning how medicine and surgery were practiced on the continent. When he returned to New York, he started a surgical practice. His claim to fame was his approach to breast cancer. He removed breasts that contained growths at a time when others were treating breast cancer with salves and whatnot. Forty-seven of the 50 women whose breast he removed remained cancer free and surgeons ultimately adopted his approach. His operation was quite radical. It involved removing a chest wall muscle and multiple lymph nodes in the arm pit. Many women survived but had deformed chests and edematous arms. His approach was the standard for more than seventy years. Then a study proved that the women who had the radical surgery didn't live longer than women who had a simple mastectomy followed by chest wall radiation, and the medical approach to cancer of the breast evolved.

In the 1880s, Halsted started experimenting with cocaine. When Johns Hopkins opened its hospital in 1889, he was hopelessly addicted. His addiction was well known, but he still became a professor of surgery and was later named the head of the department. Halstead had been a "bold aggressive surgeon" when he was young, but he was hooked on morphine during his years at Hopkins. His operations were "excruciatingly slow and meticulous." On one occasion he must have been under the influence because he stopped operating and asked an assistant to scoot over. The man had been standing on Halsted's foot for the previous half hour. Outwardly shy, Halsted was adamant about cleanliness, bleeding control, and carefully reconstructing tissue. He also had quirks like shipping "dozens of his dress shirts to Paris or London every year for laundering. He only burned white oak and hickory in his fireplace." The rubber gloves surgeons and nurses wore in his operating rooms were not intended to prevent infections but were used because one of the scrub nurses developed a rash when she immersed her hands in mercuric chloride.

In 1894, he was troubled because surgeons in the U.S. didn't have a mechanism for developing and passing knowledge to new graduates, and he created the nation's first residency program. His young doctors devised and modified operations. They taught one another and remained at the university for an average of eight years. When they left Hopkins, eleven of Halsted's seventeen chief residents started training programs in other hospitals.

One of Halsted's trainees, Harvey Cushing, became the American "father of neurosurgery." Cushing's father, grandfather, and great-grandfather were all prominent Cleveland surgeons, and Harvey was a descendent of the nation's fourth ever Supreme Court justice. As a student at Yale, he was a gymnast and was known as the person who somersaulted backwards "with a lighted cigarette in his mouth." After medical school his money and influence allowed him to tour Europe and meet and observe the world's most prominent surgeons. When he returned to the U.S., he spent four years as a Halsted surgical resident and he learned his craft. In 1914, Harvard's Peter Bent Brigham Hospital opened its doors. Cushing became the hospital's first chief of surgery and founded the country's second surgical training program.

During the First World War, Cushing operated on soldiers with severe head wounds. The brain is surrounded and confined by a hard bony skull. Space is limited, and trauma, bleeding, or an enlarging tumor can lethally increase the intracranial pressure. Cushing cut open skulls and exposed brains up to eight times a day. In the days before we had CAT scans, growths were localized by their effect on a specific site in the body. After the war, he operated at a Harvard hospital, removed two thousand brain tumors, and his mortality rate was half that of his colleagues.

Described as "a man of ambition, boundless, driving energy, and a fanatical work ethic," he also had "a penchant for self-promotion and ruthlessness." He insulted his residents and operating-room nurses for their errors and tended to blame others for his mistakes. Outside the hospital, he paid little attention to his family and had no patience for women. When his oldest son, twenty-two-year-old William, was killed in a traffic accident, Cushing learned about it from a telephone call he received in the morning. He then informed his wife, and continued to the hospital to perform his scheduled surgery.

In the early 20th century, surgeons increasingly learned how to remove tonsils, appendices, gallbladders, and cancers. Some honed their skills when they cared for the wounded in the First World War.

In 1933, a surgeon named Evarts Graham removed a lung that contained cancer and his patient survived. Graham went on to become one of the giants of his era.

Deciding at a young age that he would follow his father's path and become a doctor, Graham went to Princeton, then Chicago's Rush Medical school. He became the president of his class, wrote a number of journal articles, and performed fifty autopsies while in school. In 1907, he graduated, and when most of his classmates "became general practitioners," Graham spent five months as a surgical intern. Part of the time was spent with his dad, who was not "by any means a polished or trained surgeon."

Believing he was qualified to perform surgery, Graham joined a group of physicians in Mason City, Iowa, a town of twenty-two thousand. He was appalled when he learned that surgeons paid colleagues a portion of their fee for referring patients. He believed surgeons should not be chosen on the basis of how much they pay the referring doctor and he became embroiled in a campaign to end "fee splitting." Considered an upstart by some colleagues he became the brunt of "personal attacks." In 1916, the U.S. entered the First World War and Graham volunteered for the army and left town.

Unable to become a battlefront surgeon because his vision was "defective," Graham became an assistant surgeon general. Young and inexperienced, he was sent to Alabama and given orders to take charge of the local military hospital. When he presented his papers to the facility's medical commander, the doctor refused to step down. He "didn't give a damn about Graham's orders" and he called Graham a fly-by-night major. Graham fought back, tried to label the colonel a German sympathizer, and got the surgeon general involved. The hospital eventually became Graham's, but the facility wasn't sent to Europe until November 1918. By that time, an influenza epidemic was killing more soldiers than the war had.

While in the military, Graham became famous because of his studies on fluid that developed just outside the lungs in the pleural cavity. In soldiers with battlefield wounds, the pneumonia that developed was often caused by streptococcus. Fluid filled their pleural cavity before adhesions had formed. When the fluid was drained too soon, a vacuum was created. The heart, large blood vessels, and other structures in the middle of the thoracic cavity, the mediastinum, were sometimes dramatically yanked to the side. Structures tore and some people died. Graham learned you have to wait for scarring to occur before you withdraw the liquid.

After the war ended, Graham was chosen to be the professor of surgery at Washington University in St. Louis. The school was growing and reorganizing. In the early 1900s, the university medical school had been rated "a little better than the worst." After the First World War, it received funding from Rockefeller and Robert Brookings, a St. Louis millionaire who grew rich selling wooden spoons and bowls. Full-time professors were hired and paid. No longer needing a private practice on the side, these doctors could devote all their time to teaching and research. The concept was new, but Graham embraced it. His department developed surgical subspecialties, and the school (my school) flourished.

An avid smoker, Graham openly scolded colleagues for suggesting that cigarettes caused lung cancer. "Yes, there is a parallel between the sale of cigarettes and the incidence of cancer of the lung, but there is also a parallel between the sale of nylon stockings and the incidence of lung cancer." In 1957, shortly before he died of lung cancer, Graham publicly conceded that cigarettes were responsible for his malignancy. In 1965, his successor, Thomas Burford, a heavy smoker who operated on countless smokers with lung cancer, told a Senate committee that he did not believe cancer was caused by cigarettes. Burford's death was in part due to emphysema caused by his heavy smoking.

CHAPTER NINE

Narcotics

On October 21, 2020, the *New York Times* told us that Purdue Pharma pleaded guilty to criminal charges for opioid sales and the Justice Department announced an $8 billion settlement with the company. Members of the Sackler family will pay $225 million in civil penalties, but criminal investigations continue.

The First War on Drugs Was Fought in China

By the time I became a physician, narcotics had long been used for pain control and had an age-old disreputable history. Mentioned four thousand years ago by Aristotle's buddy Theophrastus, the extract of the poppy plant was long known as a drug that relieved pain, affected moods, and created a craving and an addiction.

The world's first war on drugs was started by the Chinese. It came about because the British developed a love for tea grown in China and the U.K. was shelling out a lot of silver for the leaves. Needing a commodity the Chinese would buy, the Brits grew poppies in India. They peddled its extract, opium, in China and it sold well. Many got addicted, and silver started flowing back to England. By the 1800s, Chinese leaders were troubled by the way the drug affected their people and were upset because so much silver was leaving the country. They forbade the British from trafficking opium, and the Brits responded by sending 16 warships. They "arrived at Guangzhou, bombarded forts, fought battles," and won the First Opium War. As spoils of their victory, the English gained control of the island of Hong Kong and access to five Chinese ports.

About the same time, thousands of miles away, the French honored Wilhelm Sertürner and called him a benefactor of humanity because he extracted "morphine crystals from tarry poppy seed juice." A successful chemist, Sertürner was born into poverty. His parents had served the local prince, and when the prince and his family died, they had no way to earn their bread. They had seven children to feed, and when their son Wilhelm was 16, they apprenticed him to the court apothecary.

He worked and learned in Paderhorn, a German town that was founded by Charlemagne in the eighth century. It bordered on the country's shortest river. In the early 1800s, plant extracts were used as medicines, but many were impure and it was difficult to know how much herb was in each dose. Wilhelm was one of many who tried to purify opium, but he was the first to succeed. He fed the crystals he developed to stray dogs and rats, and it put them to sleep. When he took "a small quantity" for a toothache, he "experienced tremendous relief." A low dose made three volunteers "happy and light headed," and a higher dose caused "confusion." He kept publishing his findings and eventually a prominent Frenchman read one of his papers and Serturner became an important chemist. By the time he received the "Benefactor of Humanity" prize, he was probably already an addict. In his later life, he "suffered from chronic depression, was severely withdrawn." He died when he was fifty-eight years old.

In 1844, Francis Rynd, a Dublin physician who hunted foxes and was in much demand at "fashionable dinner parties," invented the hollow metal needle. A decade later, a physician in Edinburgh and another in France independently invented the syringe. In the process they helped people in pain and created an instrument that would lead to countless infections and the addiction of many. In 1897, two traveling salesmen from North Carolina met at a train station in Texas and learned they had the same birthday. A decade later, now entrepreneurs, Becton and Dickinson started the first U.S. facility that produced hypodermic needles and syringes.

In the 1930s, German chemists almost accidentally discovered meperidine, Demerol, the first synthetic drug that relieved pain but was addicting.

In the subsequent century, chemists developed a number of additional synthetic narcotics. One of them was created in 1953 by a group of European chemists led by Paul Janssen of Belgium. They added chemicals to meperidine and in 1960 created fentanyl, a drug that is one hundred to two hundred times more potent than morphine. Doctors like it. I used the drug on patients when I performed colonoscopies. Owned by Johnson & Johnson, it has recently been the cause of a number of overdose deaths.

In June 1971, President Nixon declared the U.S. war on drugs. Over the subsequent decades, the Federal Drug Enforcement Agency grew to a force of more than ten thousand. In 2016, about two hundred thousand Americans were incarcerated for drug offenses, and in 2018, more than 67,000 people in the U.S. died from a narcotic overdose. That's more than 185 people a day.

In 1995 a pharmaceutical company that was physician owned aggressively marketed the narcotic OxyContin. They funded research and paid doctors to tell their colleagues that worries about opioid addiction were exaggerated. A few doctors claimed that chronic pain was being undertreated and people were needlessly suffering. Ordinary physicians who practiced fee-for-service medicine were afraid of seeming out of touch, or of losing business, and some started prescribing narcotics for long-lasting conditions like arthritis, back pain, and fibromyalgia. During OxyContin's heyday, it "reportedly generated some thirty-five billion dollars in revenue for Purdue." Then Americans started overdosing, and in less than twenty years, 750,000 Americans died.

Purdue was sued a number of times, but the corporation kept winning until a trial lawyer named Paul Hanley signed up five thousand people. He gathered evidence and "demonstrated the company had set out to perpetrate a fraud on the entire medical community." In 2006, Purdue settled the case for 75 million dollars. It pleaded guilty to criminally misbranding, and admitted marketing OxyContin "with the intent to defraud or mislead."

"Greed Is Bad"

According to economist Joseph Stiglitz, the 1920s were years of monetary euphoria. The wealthy kept growing richer. In 1929, the financial bubble burst, banks failed, people lost their savings, and there was substantial unemployment. Many blamed capitalism, and the rich feared a populist government would take their riches away. When Franklin Roosevelt was elected president, he convinced his banker friends to allow the government to establish a "wealth tax." It provided much of the money that helped the government pay for the services it provided. Those who earned more than a million dollars a year started paying the federal government 75 percent of their top dollar. During those years and for decades thereafter, doctors, nurses, and hospitals were paid for their efforts and medical care was thought of as a shared responsibility and not a wealth-producing commodity.

CHAPTER TEN

Penicillin

In 1939, Hitler attacked Poland and started the Second World War, a six-year conflict in which over 70 million people perished. During those years, penicillin, the world's second antibiotic, became available. It was discovered in the 1920s by Alexander Fleming, a professor of bacteriology at the University of London who was "not known for fastidious laboratory organization." He once had a cold and dropped mucous into a petri dish full of bacteria. After he "placed the dish amidst the clutter at his desk," he left it there forgotten for two weeks. When he came back, he noticed that his bugs were gone. He investigated and discovered lysozyme, the enzyme in tears and saliva, was responsible for their destruction. He wrote up his finding, his account was published, and other researchers learned who he was.

On another occasion, he noted that fluid coming from a penicillium mold killed the bacteria growing in the petri dishes. He named the juice penicillin, collected a small amount, and wrote an article. It was published in a medical journal and largely forgotten.

A decade later, a group of Brits got interested in Fleming's fluid. They were led by Howard Florey, an Australian-born Oxford professor, and Norman Heatley, a chemist from Suffolk, England. They were joined by Ernst Chain, a chemist and German Jew. Chain had moved to England in 1933 because the Nazis had taken over his country. He was studying antibacterial substances produced by microorganisms.

The research team got hold of the penicillium fungus and extracted enough fluid to treat a few lab animals. In May of 1940, "they infected eight mice with a fatal dose of streptococcus." Four of the creatures were also injected with penicillin. Hours later, "the untreated mice were dead and the penicillin-treated mice were still alive. Penicillin killed bacteria and—perhaps of equal importance—didn't seem to harm the animals.

The drug's spectacular possibilities were obvious. More needed to be done, but Britain was at war and Florey's research could not proceed. That's why Florey and Heatley brought some of the mold to the U.S. They convinced scientists at the agricultural research lab in Peoria, Illinois, to help them search for a penicillium mold that would produce more than a trickle of the magic juice. Mary Hunt, a lab assistant, found the super mold growing on a cantaloupe in a nearby store. It was "fifty times more potent than anything previously tested, and it became the primogenitor for almost all of the world's penicillin."

Florey decided he would not attempt to patent penicillin. He thought he and his fellow researchers "should not be concerned with monetary awards and should make penicillin available for patients." An American microbiologist who worked on the project saw things differently. His name was Andrew J. Moyer and he was the expert in molds. Working at the laboratory in Peoria, Illinois, his team "came up with the idea of increasing the yields and production rate of penicillin by culturing it in a mixture of corn steep liquor and lactose." In 1945, Moyer applied for the U.S. patent and it was granted three years later.

After the scientists isolated penicillin, they looked for a manufacturer that could generate large amounts of the antibiotic. Pfizer stepped up big. The company's engineers had learned how to ferment gluconic acid in deep tanks, and they decided to use a similar technique to make the antibiotic. The company purchased an old ice plant in Brooklyn that contained fourteen 7,500-gallon tanks, and they converted the building into a highly productive penicillin factory.

The penicillium mold wasn't the only organism that protected itself from bacteria by making a poisonous juice. In 1923, a Rutgers researcher described an actinomycete, a threadlike cross between a mold and a bacteria, which produced something that killed nearby bacteria. During the following decade and a half no one did anything about his observation. Then a former Rutgers student learned how to isolate antibacterial agents from soil microorganisms. Selman Waksman, the head of the soil bacteria department, took note and decided he would discover a better antibiotic.

He put his Ph.D. students to work and they isolated and tested juice from a number of actinomycetes. All their liquids were given to animals and were toxic. In 1943, one of Waksman's Ph.D. students isolated streptomycin from an organism in the straw compost and stable manure on the Rutgers College farm. When the chemical was given to guinea pigs that were infected with tuberculosis, the animals weren't harmed, and the antibiotic eradicated their tuberculosis. Streptomycin was produced by Merck pharmaceuticals and was later shown to help treat bubonic plague, cholera, and a number of other infections.

The Ph.D. who isolated streptomycin, Albert Schatz, had spent part of his youth on a New Jersey farm that didn't have running water, heat, or electricity. He married a woman who tolerated, or perhaps loved, dates where he "showed her how to look for slime molds and fungi." The couple later took Shatz's actinomycete cultures along when they went on their honeymoon.

At the request of his professor, Shatz gave Rutgers the U.S. and foreign patent rights to streptomycin. When he relinquished ownership he presumably thought he was contributing to the health of many. He believed streptomycin should be made inexpensively available and "that neither he nor the Rutgers foundation should or would profit from the antibiotic." Waksman, the department chief, was the second author on the paper that announced the antibiotic's creation, and he seemed to be claiming credit for the discovery. He allegedly told Schatz he did it because he was a pawn of the university and had no choice. "Schatz grew resentful and left Rutgers University bitter and penniless." When he later learned that some of the royalty money was being channeled back to Waksman's research institute, Schatz sued. The men settled out of court. In 1952, Waksman received the Nobel Prize for the discovery of streptomycin and he didn't mention Schatz by name in his acceptance speech.

In 1950, the precursor of the antibiotic Terramycin was isolated from another actinomycete, and it was produced by Pfizer.

Over the last half century, a large number of antibiotics were developed and used to treat diseases. Bacteria that developed resistance became more commonplace. Antibiotics were fed to animals to enhance their growth and to cure and prevent illness, and that played a role in the development of resistant germs.

In the decades before I entered the profession, students spent a lot of time learning about two of mankind's chronic transmissible diseases: syphilis and tuberculosis. Scientists still aren't sure if syphilis arrived in the New World in the body of one of Columbus' sailors or if the bacillus was acquired from inhabitants of the Americas. Some suspect it, along with maize and potatoes, was then brought to Europe. Some are still periodically examining the remains of ancient humans and looking for evidence.

In the years before I went to medical school, syphilis was called the great imitator. If someone got sick and syphilis was the possible cause of the problem, doctors were taught to always rule it out.

Penicillin cured the disease. By the 1950s, it wasn't gone but the number of people walking around with unsuspected disease had decreased dramatically. The VDRL, a blood test for the disease's presence, was drawn every time a person entered a hospital or obtained a marriage license. The test was invented before the First World War in a U.S. Public Health Service "venereal disease research lab." Once penicillin became available and syphilis was curable, the incidence of the disease dropped, and the blood test was no longer required in most states.

CHAPTER ELEVEN

Hormones—
Cortisone and Epinephrine

I n the late 19th and early 20th centuries, scientists isolated and learned about a number of proteins that are created in endocrine glands and travel by blood and other means. One of the hormones, insulin, helps sugar get into cells. Growth hormone makes bones grow. If a man is given estrogen, he grows breasts. If a woman is given male hormones, she grows a beard. Thyroid and adrenal hormones help regulate many of the body's metabolic functions. Some hormones are "vital." When they are totally absent, people don't function normally and some die.

Adrenal glands sit atop the kidney and produce cortisone, epinephrine (adrenaline), and a number of other hormones. In 1940, war seemed likely and the government funded cortisone research. They allegedly believed the hormone would allow pilots to fly up to forty thousand feet without oxygen.

Three people got the Nobel Prize for isolating the adrenal hormone cortisone. The first, Phillip Hench, was a Mayo Clinic rheumatologist who loved Sherlock Holmes and had "one of the more remarkable Sherlockian libraries ever assembled." He wanted to know why a woman who had severe rheumatoid arthritis improved dramatically when she was pregnant. He suspected a hormone was at work.

Hench's "Watson" was Edward Kendall, the chief of biochemistry at the Mayo Clinic. Before he came to Mayo's, Kendall had isolated thyroxine, the hormone produced by the thyroid gland. He was described by one writer as "charismatic with a generous personality." The so-called friend, also wrote that Kendall was stubborn. He told about a time when he challenged Kendall and the scientist grew sharply impatient and sarcastic. During a five-year period in the 1930s, Kendall's lab processed 900 pounds of beef adrenal glands each week. In total, they worked with 150 tons of adrenal glands and extracted 9 million dollars' worth of the hormone epinephrine for Parke-Davis. Kendall kept the adrenal material that the company didn't want. He separated and analyzed the five compounds that the gland made. Each was a steroid and they varied in potency.

During the summer months, Kendall and his family would live in a cottage on Lake Zumbro, a six hundred–acre reservoir in southern Minnesota. Kendall liked to take canoe trips with his sons and he invited members of his lab to the cabin for the Fourth of July. The running joke was the prize Kendall always offered for any guest who would cross Lake Zumbro in a tub.

In the 1940s Kendall's wife "experienced the first of a series of periodic mental illnesses," one son developed cancer and died, and another son committed suicide. I'm not sure how Kendall dealt with the losses.

In 1948, scientists at Merck synthesized a large quantity of compound E., and a year later Kendall and Hench named the hormone cortisone. Hench gave it to a woman with severe rheumatoid arthritis and watched as she dramatically improved. It took a while before doctors learned about the hormone's very significant side effects. In 1950, Kendall, Hench, and a third scientist received the Nobel Prize for cortisone. The following year, Kendall retired from the Mayo Clinic and set up a lab near Princeton, New Jersey. He spent the last two decades of his life in a nearby lab unsuccessfully trying to create something.

Cortisone dramatically changed Jack Kennedy's life. Kennedy was the 35th U.S. president and the son of a wealthy Irish immigrant. In the 1940s, doctors found he had Addison's disease—his adrenal glands didn't make enough cortisone. People with the condition are anemic, have abdominal pain, and lose weight. Some have weak muscles and get dizzy when they stand. Eighty percent of the time, disease is the result of an immune system disorder, and in 20 percent, tuberculosis is the responsible offender. As an adult, Kennedy collapsed twice: once on a congressional visit to Britain and a second time at the end of an election campaign parade. The first time, the diagnosing physician told one of Kennedy's friends: "That young American friend of yours, he hasn't got a year to live."

Another of the adrenal hormones, epinephrine, is produced by the innermost cells of the adrenal gland. A pure injectable form was first used to treat asthma more than one hundred years ago. It remained relatively cheap and available for most of the 20th century. Then, between 2007 and 2016, the list price of two injectors full of epinephrine went from $94 to $609. The 500 percent increase made the papers and a congressional committee held a hearing. The CEO of the responsible drug company, Heather Bresch, was asked to explain why she was charging so much for auto-injectors filled with epinephrine that are used for self-administration in an emergency.

Highly susceptible people use them when they develop hives, wheezing, and their blood pressure drops or they become faint. Some self-administer epinephrine when they are severely allergic to bees, have just been stung, and are starting to have physical symptoms. People take out their device, remove the top, put the needle end against their thigh, and press a button. A sharp, painless needle bursts out of the syringe, pops through their clothes and skin, and enters their thigh muscle. Then the plunger automatically pushes the drug into their body.

The auto-injector, the device that automatically squeezes in the epinephrine, was invented in the mid-1970s and the FDA approved its use in 1987. It was presumably no longer patented.

The drug epinephrine was one of the world's first hormones. Isolated from the adrenal glands of animals in the late 1800s, it was purified and patented in 1902 by Jokichi Takamine, a Japanese chemist living in the U.S., and it has long been the antidote for severe allergic reactions.

As a leader at Mylan Pharmaceuticals, Bresch was in charge of EpiPen, a contrivance her company purchased from Merck in 2007. Once they owned it, the company "pushed through legislation that made EpiPen a mainstay in schools," and they dramatically grew its price. EpiPen generated $184 million in net sales revenue in 2008, and Mylan thought they would take in $1.1 billion in 2016. That was a fivefold increase in gross income. At the time of the congressional hearings, two prefilled syringes were selling for over $600. Reporters needed to express their indignation, and they did. Elijah Cummings, the ranking Democrat, described the company's strategy. "Find an old cheap drug that has virtually no competition, raise the price over and over and over again as high as you can, and get filthy rich at the expense of our constituents." Bresch was repeatedly asked how much of the money was profit, and she kept changing the subject. The congressmen and women present that day probably assumed Bresch would quietly accept their outrage and verbal reprimand.

In at least one subsequent TV interview, Bresch, a mother of four, and the daughter of West Virginia senator Joe Manchin, appeared thoughtful and concerned. She answered questions like a well-schooled politician, with practiced talking points: Too many educational institutions did not stock EpiPens and were underserved. Her company had spent a billion dollars improving access and awareness about severe allergic reactions and how to treat them.

The publicized outrage alerted a few entrepreneurs who were watching or reading about the hearing. Some saw a way to earn a quick buck, and the FDA representative sitting next to Bresch said the agency would review new applications for epinephrine injectors within ten months. There was nothing keeping other companies from making and selling an identical product. Why not get a piece of the action?

A few companies joined the fray, and by the summer of 2017, others had commoditized the hormone. In Canada and the U.S., the price of EpiPen and another self-injecting epinephrine device, Allerject, sold for $130 a syringe. CVS Health had a deal with epinephrine syringe provider Impax Labs, and was selling their authorized generic product, Adrenaclick, for $109.99 for a two-pack. By 2011, a fourth epinephrine autoinjector, Symjepi, was being produced by San Diego's Adamis company.

CHAPTER TWELVE

Surgery of the Heart and Blood Vessels

When I graduated medical school in 1962, operating rooms and the surgeons who worked there generated a large chunk of the money that supported most hospitals, but prior to the 1950s, with a few documented exceptions, the inside of the heart and major blood vessels were off limits.

One of the "exceptions" occurred when bullets and shrapnel entered a soldier's heart during battle. During the Second World War, Dwight Harken, a battlefield surgeon, wrote his wife about a time when he used a clamp to grasp a piece of metal deep in a heart's ventricle. The pressure of the contracting heart suddenly propelled the fragment and "blood poured out in a torrent." It was like uncorking a bottle of champagne. Harken sewed the opening in the muscle closed and the patient survived. He later removed bullets and shrapnel from fifty-six wounds "in or in relation to the heart."

The son of a physician who made house calls on horseback, Harken was harsh, energetic, and aggressive. He became a Harvard professor and at one point decided to fix "rusty" mitral valves. That's the gate-like structure that opens and closes when blood flows from the atrium (the collecting chamber) to the ventricle (the muscular compartment that pumps the blood). The valve can become calcified and stiff and Harken thought he could pry the area open with a finger or cut it open with a blade. In 1948, he operated on ten patients and six of them died. Like many of the surgeons of his time, Harken kept at it, worked on his technique, and fourteen of the next fifteen patients survived.

The other notable "heart exception" involved the trunks of the major vessels. Before a child is born, a "ductus" redirects the blood that exits the right ventricle via the pulmonary artery and that (after birth) passes through the lungs. The fetal route allows oxygenated blood from the placenta to bypass the lung. When a baby is born the conduit closes. If it remains patent some of the blood that is forced into the aorta enters the pulmonary artery and circulates another time. The extra blood strains the right chambers.

A surgical resident named Richard Gross spent months operating on hundreds of dogs and thought he had learned how to safely close the connection between the major arteries. The son of a German immigrant, Gross had a congenital cataract and was blind in one eye. To compensate for his lack of depth perception, his father encouraged him to take clocks apart and put them back together. The exercise helped him develop motor skills and it "instilled a love for tinkering."

When he was the chief surgical resident at Boston Children's Hospital, Gross wanted to try to surgically close the patent ductus of a child, but his chief "forbade" human surgery. Gross waited till the man was on vacation and operated on a seven-year-old girl. The anesthesia nurse who helped him "was scared to death" because the operation had been tried at another hospital and the child had died. Gross' patient did well, but the chief of service was "furious" and, according to some accounts, Gross was fired.

In 1952, John Lewis, a surgeon at the University of Minnesota slowly lowered the temperature of a five-year-old girl who had a hole connecting the upper chambers of her heart. When the girl's body temperature fell below 86 degrees Fahrenheit, Lewis blocked blood flow to and from the organ, opened the atria, and sewed the defect shut within five minutes. During the next three years he repeated the operation on fifty additional people.

Lewis had learned about hypothermia at a 1950 medical meeting where Bill Bigelow, a Canadian Surgeon, suggested it might be possible to use cooling to "decrease the body's need for oxygen and thus allow surgeons to operate on a heart." Working in a basement room in Toronto Bigelow had controlled a dog's shivering and lowered the animal's body temperature to sixty-eight degrees Fahrenheit. After waiting fifteen minutes he warmed the animal, and it was unharmed.

In 1950, President Harry Truman called heart disease "our most challenging medical problem" and federal money started flowing. Funding for the NIH increased from $4 million in 1947 to $46 million in 1950. A number of medical schools wanted some of the money and recruited and hired talented physicians who were ambitious and wanted to make a difference. Two of the major programs that tackled congenital heart defects were located 90 miles apart in Minnesota.

Early on, each concentrated on sewing shut openings between the right and left sides of the heart. The right ventricle pumps blood into the lungs where it discards carbon dioxide and acquires oxygen. The left ventricle forces blood through the arteries of the body. Connections inside the heart cause extra blood to be pumped from the stronger left ventricle to the less forceful chambers of the right atria or ventricle. They stress the right heart, and depending on the size of the defect, kids with connections die at a relatively young age. Surgeons were unable to patch the defects in a heart that was beating and full of blood.

By the 1950s, Dr. John Gibbon had spent fifteen years trying to build a machine that could keep the body alive when the heart chambers were empty. The descendent of a long line of doctors, Gibbon was his family's sixth consecutive physician. As a college sophomore, he wanted to quit medicine and become a writer, but his father convinced him to go to medical school and use his writing to promote medical research.

When he was an intern, Gibbon cared for a young woman who broke her leg. She developed a blood clot that traveled to her lungs and killed her. He thought he could have saved her life if he had a machine that allowed blood to bypass the lungs, a "heart-lung machine," and he decided to construct one. He started building his machine when he was a Harvard doctor. His chief of surgery thought the project was nonsense but gave Gibbon lab space and a small grant. A surgeon by day, Gibbon spent his evenings "in the little room in the cellar of the Massachusetts General

Hospital." A nurse helped him. They were together every night, and eventually they married.

The apparatus Gibbon developed included a blood pump that was developed by Michael DeBakey. Its design was based on the irrigation pump that was created in the second century BC by the Sicilian Archimedes. The Gibbon heart-lung machine had to be primed with a lot of blood, was hard to sterilize, and periodically broke down. By 1942, the device was able to keep a few cats whose hearts weren't beating alive for twenty minutes. That made the papers and Gibbon was introduced to Tom Watson, head of IBM. After that, five of the company's engineers helped Gibbon modify the machine. It no longer propelled air bubbles into the circulation and it caused fewer red cells to rupture.

In 1953, Gibbon, now in Philadelphia, used his heart-lung machine and successfully closed a large defect between the left and right atria in an eighteen-year-old woman.

Building on his success, Gibbon operated on four kids with holes between the chambers of their hearts and they all died. That troubled Gibbon and he "led a movement to have Congress ban or impose a moratorium on heart-lung surgery for an indefinite time." He didn't succeed.

In the 1950s, surgeons at the Mayo Clinic and others ninety miles away at the University of Minnesota started operating on kids with congenital heart defects. Each group used a different heart-lung machine, and each center suffered through the tragic deaths of a number of children.

The University of Minnesota group was led by Walt Lillehei, a talented suburban kid whose father was a doctor. During the Second World War, Lillehei accompanied the troops that landed on the Italian coast thirty-five miles southwest of Rome. The Germans allowed the soldiers to come ashore, then mercilessly bombarded them for four months. Seven thousand Americans and Brits were killed and eighty thousand were wounded, missing, or hospitalized. Lillehei "saw the horrors of modern warfare" and it may have had an effect on the way he emotionally dealt with death.

In 1945, he was chosen to be one of the young surgeons at the University of Minnesota by Owen Wagensteen, the chief of the department. Wagensteen was the son of an immigrant farmer and had a photographic memory. He was the best student in each of his classes, and had no trouble getting accepted to medical school, but he didn't want to be a doctor. The summer before he was scheduled to start, his father exposed him to the harsh reality of farming. Wangensteen spent his days hauling manure, and at the end of the summer he decided "anything would be better than this." When fall rolled around and classes began, Owen was a medical student and excelled.

The Mayo Clinic attempt to fix heart defects was led by a young surgeon named John Kirklin. He graduated from Harvard Medical School in 1942 and was stationed in Missouri during the Second World War. Before traveling to Minnesota, Kirklin visited Gibbon, and he brought a Gibbon heart-lung machine and/or its specs with him to Minnesota. He then convinced a few engineers to help him improve it. In 1955, nine of ten dogs survived the machine. Then Kirklin used the heart-lung bypass to operate on eight kids with congenital heart problems. Four died and Kirklin analyzed his failures and learned what he was doing wrong. By the end of the year, he had closed holes between the upper chambers of the heart in twenty-nine additional kids without another death.

Fixing the gap between the two lower chambers, the ventricles, presented a special risk. The wiring that prompts the heart muscles to work as a team runs along the edge of the defect between the right and left pumping chambers. When the wiring, the bundle of His, is inadvertently tied off, the heart slows and sometimes stops beating. Surgeons who patched the ventricular defects had to learn how to avoid the wires, and most eventually did.

The CDC estimates one in a hundred children born in the U.S. has a congenital heart problem, and many undergo surgery. In 2013, Americans spent $6.1 billion for birth defect–associated hospitalizations.

In the spring of 1958, Dr. Albert Starr was running the only heart surgery team in Oregon, and was mainly performing surgery on kids with congenital defects. He was "up to his eyeballs" in clinical work when a sixty-year-old engineer with a "background in hydraulics," Miles Edwards, arranged a meeting. They talked about an artificial heart and decided instead to try to improve the function of the heart one valve at a time. It took two years before they were able to implant an artificial gate-like valve that opens to let blood through, and then closes.

Their original valve was handmade and acrylic. Later valves were made of stainless steel, and still later a non-corrosive material. By 2018, in the U.S. alone, 182,000 metal and tissue heart valves were replaced each year. By 2020, the global market for heart valve repair and replacement had grown into a $6 billion-a-year business. According to a publication called *Medgadget*, in 2014, the Cleveland Clinic was paying $4,000 to $7,000 for a valve that was replaced surgically and $32,000 for an aortic valve that was inserted through a groin artery.

Vascular Surgery

Medicine, like industry, has innovators who can mentally visualize the "possible" and find a way to make it happen. In 1964, Michael DeBakey, a Houston doctor, was, best I can tell, the first doctor to cut out a stretch of superficial vein from a person's leg and use it to create an auxiliary pipeline—a blood vessel—that bypassed a narrowing in one of the coronary arteries. DeBakey didn't report the operation to the medical community until 1973 because he wanted to be able to prove the graft had remained patent.

He once said: "You learn on animals. We did hundreds of bypasses on dogs and were fifty percent successful. The first coronary bypass was an accident. The patient was scheduled for an endarterectomy, an operation where we separate the internal lining of a diseased vessel from plaques on their surface. We'd been doing the operation since 1964." (Plaques are accumulations of lipid particles, foam cells, and debris covered by a fibrous cap. They periodically develop on the endothelium, the inner wall of blood vessel cells. When plaques are numerous or large enough, they can affect the flow of blood. When they rupture, the body's reaction can cause a heart attack or stroke.) The nature of the first patient's coronary artery blockage was such that Debakey couldn't separate the adherent particles. The man's coronary arteries were so bad that he couldn't get him off the table (and expect him to survive). So he did what we had been doing in dogs. It was the first coronary artery bypass. "Fortunately it worked. You have to take risks." The first carotid endarterectomy—cleaning the debris off the inner wall of an artery that supplies blood to the brain—was performed on a man with recurrent episodes of temporary paralysis. He had TIAs (transient ischemic attacks). He was a bus driver and surgery stopped his attacks. He lived for 19 years and died of a heart attack. The first aneurysm of the aorta in the chest was causing pain. We knew how to fix aneurysms in the abdomen and the man was hurting so he took the risk and agreed to surgery.

DeBakey virtually created vascular surgery. The son of Lebanese immigrants, he was a combat surgeon during the Second World War and "convinced the surgeon general to form what would become the mobile army surgical hospital (MASH) unit." As a youth, he liked working with his hands. He and his brother took apart car engines and reassembled them. His mother taught him to knit and sew, and "by the age of ten he could cut his own shirts from patterns and assemble them."

He later used his boyhood skills to create a vascular graft from Dacron. As he said in an interview: "When I went down to the department store . . . she said, 'we are fresh out of nylon, but we do have a new material called Dacron.' I felt it, and it looked good to me. So I bought a yard of it . . . I took this yard of Dacron cloth, I cut two sheets the width I wanted, sewed the edges on each side, and made a tube out of it. . . . We put the graft on a stent, wrapped nylon thread around it, pushed it together, and baked it. . . . After about two or three years of laboratory work (and performing experiments in dogs), I decided that it was time to put the graft in a human being. I did not have a committee to approve it. . . . In 1954, I put the first one in during surgery on an abdominal aortic aneurysm. That first patient lived, I think, for 13 years and never had any trouble." As a Baylor University surgeon, Debakey revolutionized our approach to narrowed blood vessels.

He liked to visit the patients he had recently operated on at 5:00 AM, and he was often accompanied by an entourage of more than fifty medical students, surgeons-in-training, and famous surgeons from all over the world. TV doc Gabe Mirkin recalled studying late into the night and seeing "the light in Debakey's office stay on as he wrote papers often beyond 3:00 in the morning."

Three years after his first wife died, the sixty-seven-year-old surgeon met and married an attractive thirty-three-year-old German woman "who dabbled in acting and painting." When the pastor performing the ceremony asked if Debakey preferred to sit during the ceremony, his bride said, "He stands for hours for his operations. He'll stand for this." Debakey performed over sixty thousand operations in his lifetime and became a surgeon to the stars. His patients included the Duke of Windsor and the Shah of Iran.

Analysts from *Medgadget* wrote that people are no longer going to the dime store and buying and sterilizing Dacron. Grafts produced of polyester, polyurethane, biosynthetic materials—and Dacron—are commodities that are produced by at least 10 manufacturers. They compete with one another for the $2.66 billion (2017 number) that is being spent in the global market for vascular grafts.

CHAPTER THIRTEEN

Safety: Anesthesia— Checklists—Malpractice

In the late 1800s and 1900s, washing hands and sterility made surgery and childbirth safer. During the COVID-19 pandemic, infections are prevented when people wear face masks, meet outdoors, and maintain a social distance from one another.

One hundred years to the month after William Morton introduced the world to anesthesia, I was asked to evaluate a patient who had been harmed by halothane, the inhalant most anesthetists used in the 1960s. I was the assistant chief resident at the San Francisco VA hospital, a collection of buildings near the Pacific Ocean that are often blanketed in fog, and one of the residents asked me to see a patient who was recovering from a hair transplant.

The man had been fine before surgery, but the whites of his eyes were now yellow, he had no appetite, and he was weak. The day before I saw him, an anesthesiologist had used halothane to put him to sleep. Plugs of his hair were then harvested from the back of his head, and planted up front.

The patient told me this was his second transplant. He turned yellow after the first surgery and thought he knew what was happening. When he was a soldier in the South Pacific, some of his fellow servicemen had developed hepatitis and their eyes were yellow. He feared that if his physician knew about his jaundice, the doctor would refuse to perform the second set of hair transplants. So my guy decided not to tell anyone. This happened a few years before liver transplants were being done, so people with failing livers could not be rescued. Over the next week, the patient's condition got worse, his abdomen filled with fluid, he sank into a coma, and he died.

I felt his liver disease had to be a reaction to his anesthetic. At the time, most anesthesiologists didn't believe halothane hepatitis was a real entity. The inhalant was smooth and well tolerated. Few anesthesiologists knew of a case of liver failure that couldn't be written off as a complication of a patient's underlying medical condition. I was a budding gastroenterologist, had read accounts of halothane hepatitis in the GI literature, and I believed the condition was real. Turns out that halothane causes liver failure and death in one of 35,000 patients.

Anesthesiologists wouldn't change their ways without proof, and they got it in 1969 after an M.D. anesthesiologist visited a Yale liver specialist named Gerald Klatskin. The anesthesiologist claimed his urine darkened and his skin yellowed every time he administered halothane to a patient. He had spent years learning how to anesthetize people and he thought his livelihood was being threatened by the inhalant he was using.

Klatskin decided to test the man's theory. He biopsied the anesthesiologist's liver and proved it was normal. Then the anesthesiologist inhaled halothane, his eyes and skin turned yellow, and a second biopsy showed an injured liver. Klatskin published a report of the case and U.S. anesthesiologists stopped administering halothane. It is still "widely used in developing countries.

In the last half of the 20th century, diethyl ether and later chloroform were used to anesthetize people undergoing surgery. For many, the current anesthesia of choice is propofol, the drug that killed Michael Jackson. Some operations, of course, are performed using spinal anesthesia, a nerve block, or a local infusion of lidocaine.

Ten years ago, I was writing a book about the way medicine had changed in my lifetime, and I interviewed the chief of anesthesia at a Kaiser hospital. He claimed that nowadays general anesthesia is safer than crossing the street. During the prior 50 years, while the rest of medicine was inventing new operations and trying to cure more diseases, the top anesthesia thinkers were obsessed with safety.

They had long since learned how to put a person into a state where the patient heard and saw nothing, was impervious to pain, and had muscles that were totally relaxed. The anesthesiologist credited his profession's emphasis on safety to the skyrocketing cost of malpractice insurance. I'm sure doctors in the field thought their care was excellent and wondered why they were being singled out. But the numbers said it all. In 1974, 3 percent of all American doctors who bought malpractice insurance were anesthesiologists, and these were the very doctors who were responsible for 10 percent of all malpractice payouts. Insurers seemed to have concluded that the care they provided was "below the standard."

Malpractice is not a good way to judge medical quality. Doctors are sued when something major goes wrong and when the responsible physician is arrogant or seems to be hiding something. It also is easier to sue someone you have never consciously spoken to or interacted with, someone who has never become a real person with feelings and regrets.

Nonetheless, rates were rising and something had to be done. The anesthesia societies embarked on something they called the "closed claim project." They reviewed malpractice suits that had run their course—that had been litigated, settled, or just dropped by the plaintiff. Discovering what went wrong did not create a legal or other risk for the involved doctor.

Data for events prior to 1990 revealed that, in a third of the cases, the person whose families sued had died or had suffered brain damage. Forty-five percent of the time, the harm was caused by a "respiratory event." When anesthesiologists induce coma, they become responsible for the movement of air into and out of the lungs. To make sure the vocal cords don't go into spasm, they slide a tube through the mouth and pharynx, between the vocal cords, and into the bronchus. Then they aerate the lungs and the body. In 7 percent of the "respiratory disasters," the anesthesiologist mistakenly slipped the breathing tube into the esophagus, the tube that transports food and drink to the stomach. Its entrance is located in the pharynx just above the vocal chords.

In 12 percent, intubation was difficult, and the body was deprived of air for a period of time. In another 7 percent, the doctor got the tube in the right place but didn't ventilate the lungs adequately.

Twenty-five percent of the lawsuits were the result of arrhythmias of the heart, a drop in blood pressure, and heart attacks. Nerve damage due to poor positioning and compression of nerves caused 21 percent of the problems.

Anesthesiologists sometimes instill Novocaine or alcohol into nerves in an attempt to mitigate chronic pain. When someone injected a person who was taking blood thinners, the shots sometimes precipitated bleeding, and leaking blood caused damage.

Six percent of the cases were prompted by burns caused by electrical cautery or by IV bags of fluid that were overly warmed.

There were people whose blood pressure dropped and they lost vision. Some airways had been damaged during a difficult intubation. A few people had back pain, emotional distress, or eye injuries. (Anesthesiologists work close to the eyes.)

Seventy nine percent of the problems were attributed to lack of vigilance. The specialty has an old saying: Putting someone to sleep starts with seconds of panic (intubation), and is followed by hours of boredom.

After the anesthesiologists learned what they were doing wrong, they disseminated their findings, made recommendations, and general anesthesia became safer.

Anesthesiologists and nurse anesthetists now have tools that make it possible to intubate almost everyone. Small, flexible instruments containing long fiber-optic bundles allow the anesthesiologist to see into dark corners. Some scopes have chips on their tips and send images to a TV screen. Anesthesiologists and anesthetists are able to confirm the endotracheal tube is in the right place with an ultrasound examination. They can measure and monitor the carbon dioxide level in the air that exits the lungs. If the level gets too high, ventilation may be inadequate. Complex machines regulate and check the movement of the gases. Bells ring and a beep sounds when something is amiss.

Surgical safety got a boost in 2007 when checklists were promoted by Harvard surgeon Atul Gawande. The idea of a list of procedural steps had previously been developed by the aircraft industry as a result of crashes that were caused by a minor oversight. Gawande pointed out that, during and after an operation, even surgeons who are super-specialists sometimes neglect small details. The lists pose questions like: Are antibiotics or blood available? Did we mark the breast or limb that we're going to operate on? The World Health Organization assessed their impact on surgical mortality. Between 2007 and 2008, a nineteen-item checklist was used in 3,733 surgeries in 8 hospitals in Canada, New Zealand, the U.S., England, India, Jordan, and the Philippines. The death rate fell from 1.5 to 0.8 percent, and inpatient complications decreased from 11 to 7 percent.

Do Malpractice Suits Make Health Care Safer?

"Never go to a doctor whose office plants have died."
—ERMA BOMBECK

A t eighty, he was gray, trim, and smiled a lot. His heart had been damaged by a prior heart attack, but it still pumped well enough to get him through his daily nine holes of golf. The radiologist noticed a rim of calcium had appeared on the inner wall of his aorta, the main conduit of his abdomen. It was visible on an X-ray and it revealed an aneurysm, a bulge that had formed on the artery. We followed its size with periodic "pictures." When it reached a size where the risk of rupture and sudden death outweighed the hazard of an operation, I sent him to a surgeon.

The surgery went well, and six weeks post op he was golfing. His only complaint was abdominal pain that was tolerable but strange. I ordered an X-ray and the radiologist called. Surgical sponges are marked with radio opaque threads that are easy to spot on an X-ray. The surgeon had left a sponge in the abdomen.

Replacing an aorta is bloody business. A portion of the large vessel has to be closed off at both ends before the replacement graft is sewn or stapled in. When the damaged segment is removed, blood leaks into the abdomen. The surgical team suctions some from the cavity and blots the rest with cotton diaper-like "sponges." A residual red patina covers the outside of the intestines and organs. It often becomes hard to tell a rag from normal tissue. The sponges that are thoroughly soaked are stacked in a corner of the room. Cloths are always counted before the operation starts and counted again before the abdomen is sewn shut. If a sponge is missing, the abdomen is scoured until the fabric is found. In this man's case, someone obviously counted wrong.

Large foreign materials left in the abdomen don't always cause problems, but they can get infected and their presence is a clear sign that a mistake was made and another operation is indicated. The Latin term *res ipsa loquitur*, "the thing speaks for itself," indicates a situation where lawyers don't need a witness to prove malpractice. The evidence of "wrongdoing" is obvious and irrefutable.

I called the patient and told him about the X-ray findings. When he met with the surgeon, my colleague was contrite and offered to perform another operation. In malpractice lectures, doctors are taught that if a mistake was made, admit it as soon as you know something went wrong. Accept the blame. Treat the injured party as you would like to be treated. Most errors don't lead to a lawsuit. Legal action becomes more likely when the harm is great and prolonged, or the patient or family is upset with the care they received.

My patient wasn't angry or vengeful, and he didn't want to risk a second operation. He filed a lawsuit and asked for $50,000 for his pain and suffering. Our insurance people reviewed the file, admitted guilt, and offered $10,000. They may have been low balling him, but it didn't work.

Months passed, and neither side was giving in. An expensive trial seemed imminent. Then, one day, my patient had a heart attack on the golf course and died. His family was not compensated.

During the last fifty years, medical providers have grown in their ability to improve the quantity and quality of the average person's life, and physicians and patients are taking more risks. At the same time, the price of medical care has increased and medical liability costs (as well as malpractice insurance premiums) are thought to be the source of 2.4 percent of the money we spend.

Unfortunately, at times, doctors and nurses seem to have poor people skills. We doctors may be good students, technically proficient, and methodical. But too often we are running late. Our schedules are too tight. We seem to be impatient, and—in the age of computerization—we spend too much of the visit looking at a screen.

Errors happen. The wrong medicine. The wrong dose. Bed sores. A preventable fall that leads to an injury. A sponge left in an abdomen. A cancer that should have been discovered early. Harm at the time of birth that has lifelong consequences.

The majority of physicians in a quoted survey said that when something goes wrong, they provide "only a limited or no apology, limited or no explanation. They give limited or no information about the cause." The article's author, a physician who specializes in malpractice, thinks that too often the problem is caused by a physician's need to protect his or her ego and a system that allows doctors to shift blame.

Prior to 1970, medical malpractice suits were rare. That year an estimated twelve thousand claims were filed, but a third were quickly dropped. Many who had a legitimate case did not sue, and the suits of four of five were decided in favor of the defendant. When the plaintiff won, many of the awards didn't cover the victim's litigation cost.

As the 20th century wore on, the legal rules of the road changed in one state after another. Lawyers were increasingly allowed to file claims based on errors that were so obvious that—*res ipsa loquitur*—the blunder "spoke for itself." Claims could be based on the absence of adequate informed consent. Charitable and nonprofit hospitals had once been shielded from litigation in the belief that "paying money to the victims could damage the facility's ability to treat patients." Then, in one state after another, hospitals became responsible for errors that were their fault. Massachusetts limits their financial vulnerability to $20,000.

Jury awards started fluctuating wildly. At one point, injured parties in California and Nebraska on average received awards that were twenty times higher than they were in "low-activity states, such as Maine.

In 1975, California lawmakers limited compensation for "pain and suffering" to $250,000. No limit was placed on the amount of money that could be awarded for costly medical care, lost income, or inability to earn a living because of the malpractice.

The majority of the injured don't sue. In one analysis, "approximately 70 percent of claims were closed with no payment, and defendants won the majority of cases that went to trial." People who litigate tend to have more severe injuries and there has been an increase in the number of million-dollar-plus awards.

The cost of insurance varies from state to state and it does contribute to the cost of medical care. During the last half century, there were three periods of time when malpractice insurance premiums rose dramatically and insurers left some states. When the lack of "coverage" for medical care was deemed "critical," some legislatures established patient-compensation funds and joint underwriting associations.

I was a salaried doctor in a large physician-owned group (Kaiser). During my forty years there, I was sued a few times, and it was emotionally painful. I'd prefer not to judge myself.

I don't really know what happens to physicians who practice fee-for-service medicine. Some share small offices with one or several colleagues. When sued, I assume they are forced to deal directly with representatives of an insurance company.

I never had to pay a malpractice premium. Kaiser, my employer, was big enough to be self-insured. In the event of a multimillion-dollar settlement—a severe injury that led to expensive lifelong care—an umbrella insurance policy kicked in.

I wasn't forced to deal with a representative of an insurance business or company lawyers whose chief tactic was delay and endless expensive depositions. Suits that alleged malpractice initially went to a fellow physician who was not always that sympathetic, but who usually was one of the group's brightest and best. Our malpractice doctor typically spent half of his or her time dealing with legal allegations and the other half dealing with patients. After a lawsuit arrived on his or her desk, our colleague read the chart, evaluated the case, and reached a preliminary conclusion.

All doctors who participated in the care of the injured person were always named in each lawsuit, and everyone who was being sued was informed. In serious cases, the group hired one or several outside experts and asked them to assess the case. When necessary, they got opinions from additional experts. Physicians who were paid for advising us could not become a witness or consultant for the plaintiff. If the outside experts thought we were negligent, our people tried to settle.

A large medical-legal department handled the paperwork, the release of information, and the technical matters associated with lawsuits. Skilled, knowledgeable company lawyers gave advice and guided physicians through depositions. When indicated, outside lawyers were hired to handle individual cases.

In my later years as a physician, everyone who had Kaiser health insurance signed an arbitration contract. Cases did not go to a jury. Judges were chosen who were acceptable to the lawyers from each side. I'm told the approach does not affect overall malpractice costs, but it was easier on the psyche.

When a plaintiff is awarded compensation in excess of $30,000, the state of California gets involved. If the case is settled, our group has to name the physician most responsible. That name is posted on a state website. If there was a trial and the money was awarded by a judge, the board of medical examiners gets to decide who to blame.

Doctors responsible for larger settlements often have to appear before a really tough California medical board. Most walk away with a reprimand, but about one in ten medical licenses are revoked or suspended.

Our tort system is based on blame and fault. To prevail—to win—a plaintiff has to prove that the defendant owed a duty of care, that the defendant breached the duty, and that the breach caused an injury. Plaintiffs' attorneys usually work on a contingency-fee basis, and take a percentage of the award when they win and nothing when they lose. Numerous surveys have concluded that in most cases of negligence the doctor was not sued, and when doctors are sued, the harm was usually NOT the result of negligence. Some doctors claim that the fear of malpractice leads to unnecessary testing and plays a major role in health care costs.

"Many countries—like Sweden, Finland, New Zealand, Quebec Canada, and Australia—have a no-fault system." Compensation is based on proof of "causal" connection between treatment and injury. Their structure awards damages to patients without proof of provider's fault or negligence. Their method encourages physicians to collaborate in their search for the cause of the injuries. Although the application of no-fault system differs slightly in each country, the basic idea is to eliminate fault or blame and make the claim procedure simple "so patients with meritorious cases can access the system easily." It also makes it easier to identify and fix problems

When I first came to Kaiser, the medical group had a monthly morbidity and mortality conference. Mistakes were openly discussed, problems were identified, and lessons were taught. I don't know if our discussions would have been admissible in court, but I think "legal fears" are at least part of the reasons that the conferences ended a few decades before I retired.

Doctors in the military aren't immune from patient complaints or administrative action, but the government can't be sued. In the 1940s, a serviceman died after a surgeon "left a towel in his abdomen." The family sued, and in 1950 the U.S. Supreme Court, by creating a "doctrine" called the Feres rule, decided the government was not liable. The decision was later extended to include anyone receiving medical care from "the government." The court rationalized the approach by discussing the special relationship that exists between service members and the U.S. Some judges apparently feared lawsuits might affect military discipline. The decision of the court has been challenged and additional suits have been filed, but the court has not chosen to rehear the matter.

CHAPTER FOURTEEN

Minimally Invasive Surgery

I n 1961, during my third year in medical school, I spent a week with a physician in a small town an hour's drive from St. Louis. The doctor I shadowed had removed my tonsils when I was six and was a friend of the family. He cared for common injuries, delivered babies, and performed a few operations. When someone needed complex surgery, he or she was sent to an experienced surgeon in St. Louis. Once a week, the St. Louis surgeon came to the small Illinois town's hospital. He spent the day operating. Our family friend scrubbed in and learned how to proceed once the cavity was open. Some problems couldn't wait and the local doctor wanted to be able to treat them.

Fifty-five years ago, as an intern, I lived and slept in a hospital-provided room and worked most days and every third night. Meals were free, and I was paid $55 every two weeks. When rotating through surgery, I rolled out of bed between 5:00 and 6:00 AM and accompanied the cowboys of the hospital, the general surgeons, as they made inpatient rounds. We ate breakfast and washed hands for ten minutes by the clock. Mornings were spent tugging on a retractor, stretching open the edges of the wound while a gallbladder was being removed.

Long skin incisions were usually needed when the surgeon was operating on someone who presented to the emergency room with severe pain and rigid abdominal muscles, or when merely touching the belly made the patient withdraw. Sometimes we wondered if the appendix had ruptured.

When a person had been shot or stabbed, surgeons feared blood might be pouring into the abdomen; if the X-ray showed air outside the intestines, we suspected an ulcer had perforated.

Operations started with a lengthy slash through the skin and the fat layer followed by a pause to buzz (electrically cauterize) or tie off the severed ends of bleeding arterioles. A series of additional gashes cut a path through layers of fat and muscle. A scissors sliced open the peritoneum, the membrane that surrounds the organs. The opening had to be large enough for the surgeons to see what they were doing and to get their tools and hands inside. Sliding a large light over the opening, an assistant would aim the beam down the long, dark tunnel. The smaller the gap, the more precise the ray had to be. Deep inside, tissue was cut, clamped, and sewn by skilled people using long-handled tools. Adhesions, fibrous tissue caused by prior operations, created matted intestines and made it hard to identify various organs, blood vessels, or nerves. By slipping a hand into the cavity and palpating the edges of the liver, pancreas, and other organs—by feeling for masses or abnormalities—the surgeon "explored the abdomen." Aware of "the possibility and consequences of failure," surgeons needed to be able to improvise when they had a complication or encountered something out of the ordinary. The people doing the cutting had spent many of their waking hours practicing basic techniques like sewing and tying knots, and they were adept.

In 1984, a doctor who revolutionized our surgical approach moved to Nashville "because he loved the music." His name was Eddie Joe Reddick and he was the grandson of an Arkansas soybean farmer. He went to medical school, was drafted, and while in the army became a trained surgeon. As a military doc, he learned how to manipulate a long hollow metal tube with a light on its end, a laparoscope, and he helped a colleague locate and biopsy a liver. After he was discharged, he moved to Music City and wrote country music. (His song "I'm Listening to Hank" made it to number seventy on the country music *Billboard* chart.)

As the new guy in town, he struggled financially for a while, and made a little money assisting gynecologists perform laparoscopic procedures. Gynecologists would slip the scope through a small abdominal incision and insert it into the abdomen. Then they would introduce grasping or sewing instruments through the scope or through adjacent tiny incisions. As they peered through magnifying lenses, they tied fallopian tubes shut, biopsied suspicious growths, checked out ovarian cysts, or tore through adhesions. In the early 1960s, older devices were gradually replaced by instruments that used quartz light rods, and tips on later models had video chips that transmitted images to TV monitors.

In 1988, Reddick met a Georgia surgeon named McKernan at a medical meeting, and he was intrigued by his fellow doctor's plan to remove a gallbladder using the laparoscope. A few months later, McKernan found a willing patient and would have become the first American to remove a gallbladder through a tiny hole. The stone in the "bile bag" was big, and McKernan had to make an incision to get it out. Two days later, this time in Nashville, Reddick used a laparoscope, excised a diseased gallbladder, and pulled it out through the tiny hole. He cut through tissue with a laser and clamped tubes shut with a homemade clip.

Reddick's patients got well quickly, and his complication rate was low. He had the touch, the knack, the hidden quality that makes some surgeons really good. But he didn't come from a big medical school, wasn't a professor. I suspect he was looked down on by the men who gave lectures and wrote books. At a medical meeting the following September, he gave a talk and played a three-minute video of a gallbladder being removed laparoscopically. In subsequent months, surgeons were invited to his operating room to watch and assist. Some of the surgeons at his hospital were nervous and "urged that he not be allowed to perform the procedure," but the hospital review board gave him the green light. Someone alerted the media. The *Wall Street Journal* wrote an article about the medical maverick, and ABC World News Tonight ran a story that was seen by millions. At a subsequent surgical meeting, a few instrument companies ran tapes of his surgeries in their booths, and crowds of doctors gathered around. Then Reddick and his colleague, Dr. Saye opened a training center in Nashville for doctors who had long since completed their formal training.

Within a few years, twice, then four times as many gallbladders were excised annually and surgeons learned how to use the laparoscope when they removed inflamed appendices, plugged hernias from the inside, and removed a part of the colon that contained cancer. Renamed "minimally invasive surgery," the approach was used to wrap the upper stomach around the esophagus to prevent reflux when heartburn that was caused by stomach juices didn't responded to pills.

Chest surgeons used scopes in the thorax to identify and biopsy abnormalities, to remove parts of a lung or pleura, and to perform a number of procedures on the esophagus, the swallowing tube that runs through the back of the chest. Problems that once required a long incision and spreading of ribs could often be solved with VATS—video-assisted thoracic surgery.

Using a similar scope, orthopedists entered knee joints without damaging the surrounding muscles and ligaments. They repaired torn internal ligaments and injured cushions (menisci). At times the approach was used to mend torn shoulder rotator cuffs and to patch tears in the cartilage that lines the rim of the shoulder joint—the labrum.

Back surgeons use a scope with a camera and call the surgery minimally invasive. A few years back, I had sciatica. The pain was in the left leg and was caused by pressure on the nerve that traversed the space between my fourth and fifth lumbar vertebrae. The MRI showed a cyst was causing the problem and a back surgeon inserted a scope, removed the cyst, and the pain disappeared.

Ear, nose, and throat surgeons use scopes to perform biopsies and use lasers and cautery to destroy tissue and control bleeding.

Gastroenterologists work inside the gut and colon with video scopes. Similar instruments allow ENT doctors to better evaluate vocal cord paralysis and other benign laryngeal problems. Neurosurgeons poke scopes through the back wall of a nasal sinus and remove pituitary tumors.

A company called da Vinci developed a device that had small internal robotic hands whose actions could be monitored through a three-dimensional viewer. Surgeons learned to maneuver the appendages remotely using external, space-age, knob-dial fingers. Those who survive the learning curve and master the approach claim their resections are more precise.

Many of the tasks general surgeons once performed are now carried out by interventional radiologists. Guided by the fluoroscope, they drain pockets of fluid, biopsy shadows that look like tumors, pass catheters through blood vessels, and perform complicated procedures that halt or prevent blood loss.

When blood from the intestine can't pass through a scarred or cirrhotic liver, it flows around the organ through thinned-out blood vessels and people bleed.

Some interventionists can lower the pressure by creating tunnels through livers, a procedure called TIPPS.

Video scopes allow gastroenterologists to screen colons and stomachs, to biopsy or remove abnormalities, control gastrointestinal bleeding, remove bile duct stones, remove foreign bodies, and much more.

Worldwide, about 234 million operations are performed annually. Much of the revenue of a for-profit hospital is generated by the surgeons who operate there. I was employed by a prepaid group, and all the physicians and surgeons I worked with were salaried. The hospital had ten large operating "theatres" and the OR had a calendar that was controlled by a scheduler. Each area was manned between 8:00 AM and 3:00 PM by a hundred or so nurses, technicians, and others. In the late afternoon, most of the rooms closed. The few that remained open handled the overflow, and one was available around the clock for urgent cases. During the workday, each of the designated areas was managed by a different surgical subspecialty, and had some unique equipment.

Gawande promotes the value of introducing the patient who will undergo surgery to everyone in the OR, including the person who cleans the floor and the medical student. He thinks that knowing that they are part of a team gives everyone permission to sound an alarm if they notice a problem. Before a case starts, all the people who might play a role gather around the patient and are introduced. Then a "time-out" is called. The nurse or physician in charge asks the patient their name and what they think is about to happen. When indicated, the involved breast or extremity is inked. ("In 1995, when a nurse told Florida Surgeon Rolando Sanchez he was cutting off the wrong leg, he kept going and she started to shake and cry. He felt he had gone too far. The leg he had started to remove couldn't be repaired and there was no turning back.")

A surgeon or team of surgeons usually performed all the cases in a morning or an afternoon block. If three hips were replaced that half-day, one surgical team did them all. After each operation, the room had to be cleaned and efficiently turned around. In bloody cases, like a hip replacement, soaked pads were not thrown on the floor. They were instead bagged, and the containers were tied and slipped into the hall outside the room. Following invasive surgery, when blood in the abdomen made it hard to be sure no foreign objects remained in the space, the sponges were counted before and after the operation. Post-operatively, people went to a recovery area where twenty to thirty nurses cared for them for variable periods of time. If a spinal anesthetic was used, the patient remained in the observational area until the numbness had worn off.

A few years ago, Akron General Hospital published the fee schedule for their operating rooms. Charges depended on the complexity of the intervention. The first half hour for the least complex operation was billed at $2,718. Each additional half hour added $1,100 to the bill. For the most complex procedures, the room cost was $4,935 for the first hour and $2,200 for each additional half hour.

Free-standing facilities currently perform many of the "interventions" that were once done in hospitals. They include cataract surgery, colonoscopies, knee arthroscopies, cosmetic surgery, pain management, dental, and ENT procedures. In 2019, there were 6,100 ambulatory surgical centers in the U.S., and they performed more than half of that year's 35 million operations and procedures.

Over the decades, surgeons and interventionists learned techniques, became skillful, and found niches in hospitals throughout the country. As Atul Gawande pointed out, for the vast majority, competence is not the result of a God-given skill. It's the result of the diligence of young doctors who keep practicing difficult tasks. "While the best possible care is more important than teaching novices, everyone is harmed if no one is trained." Not surprisingly, in "teaching" hospitals, "the poorest, the uninsured, and the demented" are disproportionally the responsibility of surgical trainees. Between 44,000 and 98,000 patients die in the United States each year as a result of preventable medical errors. "Adverse events in surgery account for between one-half and two-thirds of all such events in hospitals and half are potentially preventable."

Combat injuries have become expensive government funded disasters. In a recent war, teams of surgeons, nurses, and medics traveled in Humvees directly behind the troops. When a soldier was wounded, they would stop bleeding from the liver with sterile packs, staple closed bowel perforations, and wash out dirty wounds. Since blast injuries can cause blood loss from multiple locations, they would cut the abdomen open, pack areas at risk, and cover the open wound with saran wrap. When the on-site doctors had done their best—and always within two hours—the injured were flown by helicopter to a hospital less than an hour from the zone of combat.

CHAPTER FIFTEEN

Transplantation

S ome university hospitals attract customers because they study new and unproven drugs. Others become the center for operations that require a large team of skilled players, operations like organ transplantation.

We learned organ transplant was possible in 1954 when an identical twin successfully gave his brother a kidney. That's as far as it went for decades because our bodies recognized and rejected most of the organs that doctors implanted. According to Thomas Starzl, our early efforts to prevent the immune system from destroying foreign tissue, our "three drug anti-rejection regimen," "wasn't very effective or safe."

Starzl, the American who pioneered the effort to replace a failing liver with a healthy one, grew up in a small town in Iowa. The son of the town's newspaper editor, he spent his teenage free time bulldozing giant rolls of paper into the printing presses, draining oil from the machines, and hand setting words letter by letter. He went to medical school because that's what his buddies on the football team did. While attending North-western, he lived in the Chicago ghetto. He earned his keep tending the wounds and illnesses of local employees under the tutelage of a "very competent" physician. After graduating, he started his surgical training at a Johns Hopkins program that "ruthlessly" expected young men and women to be on call twenty-four hours a day, fifty-one weeks a year. Trainees received room and board but were otherwise not paid the first year and only received symbolic compensation later. At age twenty-six,

Thomas married Barbara and he needed to earn money, so he became a surgical fellow at the University of Miami. The program was new and they paid their trainees. He cared for "vast numbers of patients," became a competent surgeon, and started operating on dogs in his garage. He obtained the poor creatures from the pound and Barbara "cared for the animals." He figured out how to remove a liver without killing the canine, and he tried to transplant the organ from one dog to the next. The blood that exited the small bowel and seeped through the liver was a problem. Newly transplanted livers couldn't deal with the flow and kept failing. Starzl learned how to detour the intestinal blood around the liver. Once that problem was solved, his transplanted dogs "were normal for almost a week; then they began to reject their new liver."

In 1961, Starzl became chief of surgery at the Denver VA hospital and used prednisone and Imuran to prevent rejection of a few kidneys. He had a modicum of success, but as late as 1978, "graft survival was unsatisfactory and patient mortality high." A little too preoccupied by the rapid changes in his craft, Starzl wrote about the day in 1976 when his wife of twenty-two years casually drove him to the airport in a snowstorm. He flew to London to present a research paper and while there received "ambiguous phone calls from his family," and "knew" he could not return home. After twenty-two years of marriage, Barbara's "forbearance had run out."

Cyclosporine, the first truly effective anti-rejection drug, was developed by Sandoz, a Swiss chemical company that, in the 1800s, manufactured dyes and saccharine. In 1917, the company hired a chemistry professor and his group isolated ergot from a corn fungus and turned it into a drug used to treat migraine and to induce labor.

In 1958, the company asked employees to take a plastic bag with them when they went on vacation or business trips, and to periodically gather "soil samples that might contain unique microorganisms." They knew penicillin was a juice that a mold excreted to protect itself from bacteria, and they hoped one of their people would find the next great antibiotic.

The dirt the employees collected and analyzed didn't produce an antibiotic, but a handful of earth from the desolate highland plateau in southern Norway grew a fungus. It produced a metabolite (cyclosporine) that lessened the immune response of lymphocytes. The man who developed it, John Francois-Borel, was a company biologist. He was a reluctant scientist who originally wanted to make art but became a researcher because "you know how art pays. I am not the Bohemian type."

In 1976, Borel told the British Society of Immunology about his drug. "A small stocky surgeon with a mop of curly black hair (Starzl's description) asked for samples." His name was Sir Roy Calne and he had been working in transplantation since 1959. After he received the "fungus juice," Calne used it to try to prevent the destruction of organs transplanted in rats and dogs, and the drug was dramatically effective.

By 1973, the Sandoz supply of fungus-derived cyclosporine was largely depleted. Large sums of money (around $250 million) would be needed to create more, evaluate its anti-rejection potential, develop a drug, and obtain approval from the FDA. There wasn't much of an organ transplant market, and the investment didn't make much sense. But with Borel's help, Calne presented his findings to decision makers at Sandoz and the pharmaceutical company agreed that the drug "looked more promising."

In the early 1980s, Starzl used cyclosporine successfully on liver transplant recipients. With his results in hand, the FDA fast-tracked approval of the medication, and in 1983 it became available for use in the U.S. It's currently made generically by a number of countries and its wholesale price is not outrageous. It is priced at $107 a month in the developing world, 121 pounds per month in the United Kingdom, and about $173 per month in the U.S. (if generic drugs are prescribed).

The second major, now widely used, anti-rejection drug was tacrolimus (Prograf). Originally isolated from the "fermented broth of a Streptomyces bacterium," it was discovered and developed in the 1980s by Japanese chemists screening "natural substances in the soil for their anti-cancer and anti-rejection properties." They performed their studies at Fujisawa Pharmaceutical, a company located at the foot of Mount Tsukuba, a green oasis with hiking trails, Shinto shrines, and a good view of Mt. Fuji.

English scientists tried tacrolimus on dogs and "declared it too hazardous to test in humans." Dr Starzl found the drug kept transplanted organs alive in some animals and "rescued some organs that, despite cyclosporine, were being rejected." Additional clinical trials "suggested that tacrolimus might be safer and better tolerated than cyclosporine."

In renal transplant recipients, the drug led to improved graft and patient survival, and that led to its routine use in U.S. renal and pancreas transplant recipients. The FDA made it official in 1994.

Fujisawa merged with Astellas. The year before Prograf (its brand name) had a generic competitor, Astrellas sold up to $2.1 billion dollars' worth of the medication.

Once effective anti-rejection drugs were available, transplanted livers survived. Starzl moved to the University of Pittsburgh and brought liver transplantation east. Three of the first four people he transplanted died and "fifty-four residents and interns in the Department of medicine signed a resolution denouncing liver transplants as unrealistic and potentially unethical." In response, Starzl admitted patients with liver cirrhosis to the surgery service. (He couldn't make the patients live longer, but he could get the internists off his back.) Nineteen of the next twenty-two transplanted people survived, and Starzl and others turned the Pittsburgh transplant program into the largest in the world.

When the FDA approved cyclosporine in 1983, a skilled (unnamed) California surgeon attempted a liver transplant and his patient bled to death. In an effort to save the person's life, he was aggressively transfused and (hearsay) the blood supply of Southern California was threatened.

Before the University of California at San Francisco started transplanting livers, they realized it was going to be an expensive commodity and they agreed to make a major commitment to its support. In addition to having access to a wide array of subspecialists, the hospital needed kidney dialysis capability, respiratory therapy support, and an extensive blood bank.

Surgeons had to be trained specifically for liver grafting and a team had to be available to recover donor organs, keep them alive, and transport them quickly. Once transplant centers knew a liver was on the way, the program would hospitalize two potential recipients whose blood type was the same as the liver. One would be the person at the top of their list and the other would be a backup, just in case. People with active infections could not be transplanted until the infection was gone.

In 1990, having spent most of his life transplanting organs and teaching others, Starzl had a heart attack and wrote a memoir. In it he mused that every person who receives someone else's organ starts seeing the world in a different way, and that medicine's ability to save a life by transplanting an organ is a legitimate miracle.

A few years before cyclosporine became available, Christiaan Barnard in South Africa and Norman Shumway at Stanford each transplanted a human heart. Neither recipient survived for three weeks. In 1971, *Life* magazine's story of an "era of medical failure" told readers that, subsequent to the first two, "166 heart transplants were performed and 143 of the recipients died."

Christiaan Barnard, the surgeon who performed the planet's first heart transplant, was born in a sheep farming region of South Africa and was the son of a pastor and a church organist. As a student at the University of Cape Town, he had little money. He was on scholarship, and had to walk five miles to school each day. After he graduated from medical school, Barnard married, had two children, and practiced medicine for two years. Then, deciding he wanted more from life, he accepted a scholarship to the University of Minnesota. He spent thirty months (many without his family) working with some of the first surgeons who repaired heart defects in children. He watched the surgeons operate and learned techniques. He often managed the machine that oxygenated the bodies of the children whose hearts were not beating. After he returned to South Africa, Barnard and his brother, who was also a surgeon, operated on forty-eight dogs, and they learned how to transplant a heart. Then he was introduced to a fifty-three-year-old man who was bedridden, had severe heart disease, and was ready to die. The heart Barnard gave him came from the body of a twenty-five-year-old woman who was brain dead as the result of a traffic accident. The man who received the woman's heart survived surgery, and the operation became front-page news. The transplanted heart worked for eighteen days, then the patient developed pneumonia, and died. Reflecting on the man's transplant decision, Barnard later wrote, "For a dying man it is not a difficult decision because he knows he is at the end. If a lion chases you to the bank of a river filled with crocodiles, you will leap into the water convinced you have a chance to swim to the other side. But you would never accept such odds if there were no lion." A second heart transplant recipient lived eighteen months.

Barnard became a celebrity, let his hair grow long, started wearing suits made by an Italian tailor, dated movie stars, and ended his first marriage. During his life, he performed seventy-five more heart transplants, created a tissue heart valve, and was married two more times. His rheumatoid arthritis eventually crippled his hands, and when he was sixty-one, he stopped operating. By 2001, the year Barnard died, doctors in the U.S. were performing 2,400 transplants each year. Eighty-seven percent lived for at least a year and 75 percent more than five years.

A month after Barnard performed the first heart transplant, Norman Shumway, a California surgeon, transplanted the planet's second human heart. At the time, Shumway had been transplanting dog hearts for ten years and knew what to do technically, but anti-rejection medications weren't available. His patient died within three weeks.

A member of the high school debate team in Kalamazoo, Michigan, Shumway originally planned to go to law school, but during the Second World War he was drafted. When the government decided they needed more doctors and dentists, Shumway was one of the soldiers tested. He scored high and chose medicine over dentistry. Assigned to a group of men who received pre-medical training at Baylor University, he was a hospital orderly for six months. Then he went to medical school at Vanderbilt. During the Korean War, he was an air force doctor. When his service ended, he joined the observers at the University of Minnesota who were learning how to correct congenital heart defects. Unable to get much hands-on training, he decided to go into private practice and joined an older doctor. The partnership was not a fit. Someone convinced Shumway to come to Stanford University, an institution that didn't have any doctors with heart surgery experience. A modest man who was relieved when someone else performed the first heart transplant, Shumway became the chief of cardiothoracic surgery at Stanford in 1965.

In 1983, after the FDA sanctified the use of cyclosporine, the transplant scene changed. In the subsequent decades, more than 700,000 people in this country lived part of their lives with someone else's liver, kidney, or heart. Kidneys, on average, lasted twelve to fifteen years; livers had a shorter life span. That's going to change now that hepatitis C (a frequent cause of liver destruction) can almost always be cured. When a person with chronic hepatitis C was transplanted, the new liver always became infected, and the organ had a relatively brief life span.

Currently, in addition to brain death, patients who have severe brain injuries but who are not "brain dead" can become organ donors if the patient consents by means of an advance directive, or if the patient's family decides that life support should be withdrawn. "To avoid obvious conflicts of interest, neither the surgeon who recovers the organs nor any other personnel involved in transplantation can participate in end-of-life decision or the declaration of death."

Some countries have a system where an appropriate dead person's organs can be transplanted into another unless the person explicitly objected while he or she was alive and competent. The U.S. and a number of other countries require specific consent. In 2015, China officially stopped using death row inmates as an organ source and began to exclusively use "organs that were voluntarily donated by civilians."

When someone "dies" and donates their organs, teams of doctors come to their hospital. Livers and kidneys are removed without dissection, and without traumatizing blood vessels. The organs are cooled and transported (sometimes by plane or helicopter) to a hospital where surgeons and recipients are waiting.

Once outside the body, "the heart is most sensitive to lack of blood flow and needs to be planted in a body within four hours." Lungs, with appropriate cooling, remain viable for six to eight hours," livers are viable for twelve hours, and kidneys for twenty-four to thirty-six hours.

A private nonprofit organ transplant organization, UNOS (United Network for Organ Sharing), "oversees" all organ procurement and transplant programs in the U.S. and makes the rules about who can do transplants and how organs are to be allocated (given) to patients."

Transplantation is a commodity that "generates significant immediate and downstream revenue." In many areas there is more than one institution that performs the service and there is "competition for referrals from medical groups and for the available organs. Larger waiting lists statistically seem to draw more organs to a center. That causes some institutions to aggressively increase the number of patients on their list." In the process, they sign people up who aren't great candidates.

Once they are listed, the people on dialysis get in line and wait their turn. Managing the list can be tricky, and my former employer did a bad job when they opened their own renal transplant center in 2004. They mismanaged the transfer of their patients' records and had to close the unit and pay a fine.

People on the liver transplant list whose blood type is the same as the organ and who are closest to death are given the first available liver. There are rules that exclude people who are addicts, alcoholics, or obese.

One day I was asked to see a middle-aged female who drove a forklift in a warehouse. She had never before been sick, was muscular, and didn't drink or have hepatitis. But her liver was full of fat and her body was swollen. She had developed hepato-renal syndrome, an uncommon disorder where the disease in the liver causes a person's kidneys to stop working. Dialysis doesn't help, and people don't live long. To survive, she needed a liver transplant. I explained the problem to the patient and with her permission I called the transplant intake doctor at the university. It was a bad hour of the week—Friday afternoon—time to go home. The patient was retaining so much fluid that her BMI (weight) was technically over the limit, and the university doctor turned the patient down because she weighed too much. I explained that part of the weight was caused by fluid retention. Her dry weight—BMI (body mass index)—didn't exceed the threshold.

"No can do," the university physician explained.

I told the patient. She cried and her sister who was in her room cried. The sister offered to donate part of her liver, but the patient refused. The woman remained in the hospital and was given infusions of a few drugs. The medications helped a little and her kidney function stabilized. When the following Friday arrived, she was still alive and on a whim I again called the hepatologist on call at the university. Each of their doctors is on call for a week, then a new physician took over. I explained the situation to the new intake doctor and she said, "No problem. Send her over." The nurse called an ambulance and the woman got a new liver and she did well.

In 2011, an average transplant of one kidney had a price tag of $260,000. Combined heart and lung transplants were costing $1.2 million. The first 180 days of post-transplant medications was costing $18,000 to $30,000.

A number of generic immunosuppressive drugs have been marketed. "Cellcept was approved in 1995. Mycophenolate mofetil became available in 2008, tacrolimus in 2009, and sirolimus in 2014." The drugs are commodities and are aggressively marketed.

It's claimed that the annual cost of U.S. transplant immunosuppressive therapy averages $10,000 to $14,000. If true, then in 2016 the 33,000 transplant recipients were (directly or indirectly) paying $330 million to $462 million a year. That becomes $3.3 billion to $4.6 billion over ten years if drug prices don't rise or fall.

In India (where the culture surrounding pharmaceutical prices is quite different from ours), the amount paid for tacrolimus was slashed 65 percent in 2016. The average recipient now pays $235 to $314 a month for anti-rejection medications.

On October 30, 1972, kidney transplant became a Medicare "right" and the government started giving the person who received a kidney transplant three years' worth of anti-rejection drugs. At the end of three years, the patient was removed from Medicare and had to pay for their own anti-rejection drugs. At that point, 22 percent of people stop taking their anti-rejection medications because of side effects, high cost, or for other reasons. When a kidney transplant recipient stops their immunosuppressive drugs, they almost always reject their kidney, and they are forced to go back on dialysis or die. This is not theoretical. It happens. And it's a problem.

A few years back, our former neighbor's son, learning he was a "match," decided to donate half his liver to an uncle he didn't know that well. His mother was a mess. Donors are screened. They must be young and healthy. But, because of the amount of donor liver needed, the operation usually carries an extra risk. The liver has a large and a small lobe. The small lobe easily replaces the nonfunctioning liver of a small child. An adult needs part of the large lobe. The donor's liver has to be "split" and a significant amount of tissue has to be removed. Over time, livers grow back, but it takes months. If too much of the organ is removed, the donor is in trouble. There are occasional complications, mainly bile leaks, and one in two hundred donors dies. There are only a few centers in the country that do at least 100 living donor liver transplants a year. The young man's mother was pretty cool and has a strong social conscience, but this was hard. Bottom line: He donated and survived. Mother and son are doing well.

In the U.S. in 2019, 23,401 kidneys were transplanted, as well as 8,896 livers, 3,551 hearts, and 2,714 lungs. Of the donors, 11,900 were brain or heart dead, and 7,397 of the organs came from living donors. In 2019, more than 112,000 Americans were on one of many transplant lists, and the wait for a deceased donor is often five or more years.

CHAPTER SIXTEEN

Taxpayer-Funded Research Is Privatized—Bayh-Dole

I n the years that followed the Second World War, the federal government paid for a lot of the research that was performed in universities and the National Institutes of Health. When a drug was developed, it was the public domain and belonged to all comers.

By the 1970s, Germany and Japan, the countries that the U.S. and the Allies defeated in the Second World War, had become competitive and America was losing status.

In 1973, Israel and Egypt fought a war. The countries that supported Egypt, the Organization of Petroleum Exporting Countries (OPEC), cut oil production. They banned petroleum exports to the nations that supported Israel. The Japanese were selling more than 20 percent of all new cars. In the late 1970s, the U.S. inflation rate and the federal interest rate for borrowing money both exceeded nine percent.

Near the end of the decade, Birch Bayh, a Democratic senator from Indiana, was "lobbied" by a professor from Purdue University. He was told something like: The economy is "in the doldrums." Companies that produced autos, steel, and household appliances were moving their factories overseas and privatization of government research could stimulate the economy. That year, Bayh and Bob Dole, a Republican senator from Kansas, proposed a law that authorized the "licensing"—in essence, the sale—of public research to private entities. The law was opposed by Admiral Rickover, "the father of the Nuclear Navy." He believed products created with government funds should be freely available. His experience in the nuclear industry showed that patents were not critical to development.

After the 1980 U.S. elections and before a new group of legislators took over, Congress passed the Bayh-Dole Act, and President Jimmy Carter signed it. It authorized universities to patent medications that were created with federal funding and to sell the license to private investors. The new titleholder would then be able to manufacture the product or sell it to the highest bidder. The law didn't say the taxpayers who paid for much of the research should get a price break when the drug came to market—and they don't. Someone must have anticipated high prices because the Senate debated a windfall profit tax—and it didn't become part of the final bill.

In 2002, one of the law's cheerleaders claimed it had "helped to reverse America's precipitous slide into industrial irrelevance." He claimed it led to the creation of "260,000 new jobs and contributed $40 billion annually to the American economy."

I first encountered the law's effect when I met a young man on a launch that was speeding through the dark waters of the Grand Canal of Venice. He explained he finally had enough money to bring his wife and child to Europe, thanks to a bonus he received. He had recently acquired the rights to a new drug for the pharmaceutical company he worked for. Called enzalutamide, it was a treatment for prostate cancer, and it was going to be big.

Prostate cancer, one of the Western world's common lethal malignancies, was found in almost a quarter of a million American prostates in 2018, and it killed 33,000 people. If it grows in bones, it can be quite painful, and when it's advanced and widespread, it's not curable.

Located between the bladder and the penis, a young man's prostate is the size of a walnut. It grows as men age, and it eventually becomes large enough to slow or obstruct the flow of urine. Now and then a mutated cell reproduces more rapidly, lives longer, and its offspring become a benign tumor. Over time, there can be additional mutations and the growth can become a cancer. Its size and spread can be slowed or halted for a period of time by interfering with the hormone that fans the fire, by eliminating testosterone.

In the 1940s, University of Chicago physician Charles Huggins showed that metastatic prostate cancer could be controlled for a few years with surgical castration or female hormones. During the subsequent decades, orchiectomy, removing the testicles, commonly kept the cancer from growing for a period of time. Then the malignancy typically started expanding.

In recent years, physicians have fought the malignancy with drugs that antagonize testosterone. When the medications stop working, the disease becomes aggressive. The growth is usually stimulated when androgens bind to cancer cell proteins called androgen receptors (ARs). That's what researchers at UCLA and Sloan Kettering tried to block. They were funded by the government and people who donate money to prostate cancer research. The medical teams spent years before they developed a drug that inhibited the actions of the cancer cell's AR. Starting with a protein that was known to have "a high affinity for the receptor, they spent years chemically altering it." (Like—take a dress pattern and add one pocket or two pockets; a zipper or buttons.) They added carbons, hydrogens, and so on, and came up with 200 candidate molecules. They tested them in the lab, using "human prostate cancer cells that had been engineered to express increased levels of the receptor."

Two of the two hundred potential drugs seemed promising. Well absorbed and not toxic, they were effective blockers. UCLA patented the chemicals in 2006 and tested them on mice. They worked—stopping mouse prostate cancer from growing and spreading.

In 2005, Medivation, a San Francisco–based "Biopharmaceutical Company," somehow learned about the drug. Signing a license with UCLA, they walked away with a majority of the patent rights. In return they agreed to fund all costs associated with the development and commercialization of MCV3100 (enzalutamide). The next big study was probably not funded by Medivation. It was performed in 2009 by the U.S. Department of Defense, and it showed that MCV3100 had "significant antitumor activity."

By October of 2009, the ownership of enzalutamide was a commodity that was repeatedly bought and sold. Medivation made a deal with

Astellas, a large Japanese pharmaceutical company. Medivation received $655 million, Astellas got "global rights" to the drug, and the two companies financed a huge international assessment. Sixteen hundred men with metastatic disease got either enzalutamide or a placebo. The men who took the medication on average lived five months longer than those receiving placebo. Treated patients had "a 37 percent reduction in the risk of death."

The FDA approved the use of enzalutamide in men who had failed standard chemotherapy. Its initial planned price was $7,450 a month—$59,000 for eight cycles—$89,000 a year.

In 2014, based on a new study, the FDA approved the use of enzalutamide as the first drug that was given to people with metastatic disease. Patients didn't have to first fail treatment with something else. The new indication meant patients would live longer after they started therapy. They would ingest more pills and buy more medicine. A year of therapy in the U.S. would cost $129,000.

Astellas had international rights and sold the medicine for a lot less in other countries. A forty mg pill, for example, was sold in the U.S. for $88. Medicare paid $69. And the price for the same product in Canada, France, and the U.K. was $20, $27, and $36. In the two years between 2012 and 2014, Medicare's enzalutamide cost went from $35 million to over $440 million annually. In 2016, UCLA sold their residual rights to Royalty Pharma for $1.14 billion, and settled for an up-front cash payment of $520 million.

In 2015, Astellas sold $2.2 billion of the drug. The following year, Pfizer bought Medivation for $14 billion, and in the first quarter of 2017 sold $131 million worth of the medication.

Thanks to the 1980 legislation, the drug that was developed with public and donated money became a commodity that enriched a few and sold for a pretty penny—a price that was usually paid by a needy taxpayer's private or public insurance.

Bernie Sanders claims that, in 2014, nearly one in five Americans between the ages of nineteen and sixty-four—35 million people—decided NOT to fill their prescriptions because the drugs cost too much.

Enzalutamide's first competitor, abiraterone (Zytiga), was never owned or paid for by the government, but its creation was financed by donated money. It was developed at the Cancer Research U.K., a charitable fund with its own research institute. In 2012, an anonymous donor gave the organization 10 million pounds ($13 million). He or she asserted that "if you do what you've always done, you'll get what you've always got." The institute promotes scientists who "think differently." It finances "the work of more than four thousand researchers, doctors and nurses throughout the U.K., and supports more than two hundred clinical trials and cancer related studies."

The prostate cancer drug its scientists created, abiraterone (Zytiga), was initially pricey and not really affordable to a guy without good insurance. Here again, Pharma was brought in after the medication was created and was ready to be tested on humans. Once more, the enemy was testosterone. Researchers wondered if the prostate cancer cells grew because they lost their dependence on male hormones—or if they were responding to testosterone made in the adrenal gland. What would happen, they asked, if a drug blocked the gland's ability to make male hormones—androgens?

When the adrenal gland makes cortisone, it also makes a small amount of male sex hormones. (Raisins and wine are both made from grapes.) Both hormones use an enzyme, CPY17, for their development, and their creation can be blocked by the antifungal agent ketoconazole. All this was known. Ketoconazole is toxic and can't be used chronically for patients with prostate cancer, but investigators were able to modify it. Using three-dimensional models, a U.K. team found a compound that wasn't toxic and "specifically and irreversibly" blocked CYP17.

They filed a patent and licensed the drug to a German pharma company, Boehringer Ingleman. Phase one studies showed the drug blocked androgen and cortisone production, but the pharmaceutical company's scientists believed that late-stage prostate cancer no longer needed male hormones to grow. They decided they didn't want to spend money on a lost cause, and Boehringer returned the drug's license. When studies proved the drug helped cancer patients, the IRC next assigned commercialization rights to publicly traded BTG, a U.K.-based health care company. BTG treated the drug as a commodity, licensed it to Cougar Biotechnology, and Cougar "began to develop a commercial product." In May 2009, Cougar was acquired by Johnson & Johnson for about $1 billion. Two years later, the FDA approved arbiraterone's use in combination with prednisone—a form of cortisone. (When it blocked the body's production of androgens, abiraterone also blocked the body's ability to produce cortisone, a hormone the body needs.)

Abiraterone (Zytiga) was U.S. approved for use as a treatment for late-stage prostate cancer in men who had already received standard chemotherapy, and it initially sold for $5,000 a month. After a mean of eight months, the drug stopped working or the average patient died—thus the cost of treating a person was averaging about $40,000.

In the U.K., where it was marketed by Janssen, its initial price was 2,930 pounds—$3,820—a month, a price that British regulators (NICE) decided was not cost effective. The National Health Service wouldn't pay and was willing to walk away, so Janssen negotiated. The U.K. got a "deal." The NHS subsequently paid 2,300 pounds ($3,000) a month "for the first ten months of therapy. For people who remain on treatment for more than 10 months, Janssen agreed to rebate the drug cost of abiraterone from the eleventh month until the end of treatment."

CHAPTER SEVENTEEN

Medical Devices

A friend awoke with chest pain and a cold sweat, and his daughter called 911. The EKG (electrocardiogram) in the ambulance showed evidence of an acute myocardial infarction. Part of the heart muscle was dying. The EKG device the paramedics attached to his chest, arms, and legs was developed in the late 19th century. It detects and records the flow of current as it travels through the heart and tells each chamber when to contract. Its "wave" is different when a person is having a heart attack. Like hieroglyphics on an ancient tomb, the squiggles were little more than nonsense before Willem Einthoven, a professor of physiology at the University of Utrecht, learned how to interpret their meaning.

The paramedic radioed ahead, and the vehicle sped to a nearby hospital where a catheterization team was available twenty-four hours a day. A half hour after they arrived, a sedative had calmed the man's brain and a cardiologist was advancing device one, a narrow, rigid, yet flexible tube that he had inserted through a groin artery. Under fluoroscopy, the physician maneuvered the tip through the aorta—the body's central blood vessel—and quickly reached the coronary arteries—the vessels that drape over the surface of the heart and deliver oxygen to its muscles. Their entrance is located just beyond the point where the aorta attaches to the heart.

As we age, smoke, and eat rich food, plaques develop on the inner wall of many of the arteries that supply oxygen and nutrition to our bodies. These "barnacles" are full of fat and foam cells. If their fibrous cap ruptures, the body tries to keep the contents from escaping by heaping platelets and clotting proteins on the exposed gap. When the pile is large, it acts like a jackknifed eighteen-wheeler and can close down the highway—in this case, obstruct the flow of blood through a coronary artery. Dye was injected into the arteries and the occluded vessel was identified. The device was passed through the occlusion and a strong balloon near its tip was inflated. It forced the narrow area open and blood flowed and oxygenated the heart. A second device, a "stent," a thin mesh hollow tube made of stainless steel and cobalt-chrome, was advanced to and through the vessel. The outer surface of the device was coated with a polymer. It "carried" a drug that was slowly released and helped preserve the stent's patency.

A cousin dodged death. Her surgeon successfully clipped the bleeding brain aneurysm, the thin-walled balloon-like bulge in the wall of one of her arteries. A few days later, the cerebral spinal fluid that normally coats the outside of the brain wasn't flowing and tension was building in her head. Her physician drilled a hole in her skull and inserted a tube into the liquid-filled chamber in the center of the brain, the ventricle. He tunneled the other end of the tube under her skin and inserted it into her abdomen. Spinal fluid poured out of the brain, and the pressure that was squeezing the brain abated. The shunt, the small hose that relieved the pressure, was created by an engineer in 1955. He was the father of Casey Holter, an infant who had hydrocephalus, too much fluid and pressure on the brain. Casey's condition troubled his doctors but they avoided the father because they didn't want to deal with his reaction when they told him that his son was going to die. John Holter, the father, was thirty-five, had been in Europe during the Second World War, and was working in a hydraulics research laboratory. When his only child was born with multiple congenital defects, he painfully watched the infant undergo a number of operations. Then Casey's head started enlarging. Liquid accumulated in the brain chamber, the pressure inside the skull increased, and the doctors didn't have a good way to treat it. To decompress the situation, according to his troubled father, Casey was taken to "the torture chamber." A doctor would insert a needle through his fontanel, the soft area on the top of his head where the skull bones had not yet fused. M.D.s would use a syringe and withdraw fluid to lower the pressure.

When a doctor finally talked to dad, he explained that if they tried to drain the fluid by inserting a tube into the skull, it wouldn't last long. The other end would be slipped into a vein, and every time the child coughed or sneezed, blood would flow up the lumen. Sooner or later, a clot would form. Holter pondered the problem. He knew autos had pressure release valves and used a ball that was displaced when the tension was high. In people, a similar approach would fail when the head was in certain positions. Then Holter thought about the nipple on a baby bottle. It opened when the baby sucked, then it closed. Using it as a model, he designed a pressure-sensitive valve. In place of a tip, it had a "slit similar to the one on the nipple." He hooked the valve to a rubber and later a plastic tube and his device worked. Unfortunately, it lost its shape when the tubing was heat sterilized. Then Holter learned that Dow made temperature-resistant silicone tubing. It didn't take him long to sort out the details, and he quickly created a tension-sensitive tube.

In three weeks, Holter solved a dilemma that had plagued doctors for more than a century, and his invention "is still widely used." Sadly for Casey, the infant had a cardiac arrest before the doctors were able to put a device in his head. He survived but his heart had stopped beating for thirty minutes and his brain was damaged. With the shunt in place, his very limited body survived for five years.

At 96, a friend's mother was clear minded and living alone when she fell and couldn't get up. Four hours later, help arrived. In the hospital, physicians diagnosed and successfully treated sepsis, which was caused by a urinary tract infection. But when they examined her heart, they heard a loud woosh, a heart murmur. A cardiac ultrasound revealed severe aortic stenosis. The valve that swings open and closed when the heart contracts, the gate that allows blood to flow out of the heart and into the aorta, had grown quite stiff and her days were numbered. The valve could be surgically replaced, but she was old and frail and open-heart surgery would be quite risky. But there now was another option. A recently approved device, a replacement aortic valve, could be inserted through a groin artery and slipped up the aorta and through the old stiffened valve. Once in place, it would, umbrella-like, open and close each time the heart contracted. People in Europe had been using the device for five years and learned that the new valves were as successful as the valves that were planted surgically. She opted for the new gadget and did well.

My wife's cousin was in her eighties and lived on the second floor. Her knees were arthritic, and the pain of going up and down the stairs had become so intense that she seldom left home. She had heard "horror stories" about knee joint replacement, but dreading the thought of a nursing home, she bit the bullet. Over the next year and a half, each of her knee joints was replaced. The arthritic surface, the eroded area on the end of the bone, was sliced off and the raw area was "re-soled." The day after each operation, she was able to walk. A few months after the second procedure, she felt normal, mobile, like she was ten years younger.

The first really successful joint replacements were performed by a talented surgeon named John Charnley. A Brit with "febrile inventiveness and a powerful command of the English language," he helped care for the soldiers who were evacuated from Dunkirk, and spent most of the Second World War as a medical officer in Egypt. In 1960, having established himself as an orthopedic innovator, he turned an old TB hospital into a hip center. A local medical equipment manufacturer, Thackeray's, fashioned the metal ball on a stem that went into the femur and the polyethylene socket. Charnley used acrylic bone cement as grout, not glue, and successfully produced a low-friction prosthesis. When it was implanted, people with bad arthritis could walk without pain. His achievement inspired a new approach to worn-out joints. According to Arthritis U.K., we've now gotten to a point where 80 percent of "cemented hips should last for twenty years." When or if they fail and a person is healthy enough, repeat surgery is usually successful. "The results are less good after each revision, but 80 percent of re-dos are good for ten years."

While Charnley was developing his low-friction hip, surgeons in various parts of the world tried to design workable knee replacements. People with disabling pain were willing guinea pigs. In 1968, the "first" total knee replacement in the U.S. was a failure. Six years later, John Insall, two fellow doctors, and engineer Pete Walker designed and implanted a prosthetic knee that allowed a patient to move naturally and without pain. Doctors and engineers were impressed, and over the last three decades, the design and procedure was periodically modified and improved by Insall and others.

Insall was born in Bornemouth, a town on the English coast that boasts "seven miles of sandy beach and an exceptionally warm microclimate." He went to medical school because too many family members were already in the military and police. After he graduated, he worked in India, and before he returned to the U.K., he wrote hip transplant pioneer Charnley and asked for "an appointment as a House officer." Charnley wrote back, "India needs doctors to treat fractures and tumors, not hip replacements." In 1963, war broke out between India and China and Insall left the country with eight dollars in his pocket. After two years he moved to New York City, and a few years later was a major driver behind the development of four knee implants. Currently, more than 700,000 American knees are restored annually. Four manufacturers make the majority of the implants and some think that by 2030, over 3 million Americans per year will get a new knee.

There currently are nineteen device manufacturers whose annual revenue exceeds $5 billion. When one firm has a patent on a gadget that is new or unique or if the device is only made by one or two companies, hospital chains pay list price. When multiple gadgets effectively do the same thing, salespeople try to "sell" the qualitative superiority of their gadget, and the amount chains pay is negotiable. Hospitals and salespeople use the list price as a starting point and reach an agreement. The contracts they sign often include nondisclosure language that forbids hospitals from telling other hospitals what they paid. If the price "leaks," the supplier can rescind the agreement and revert to list price.

In 1895, a German researcher named Wilhelm Röntgen "noticed that when electric current was flowing through his Crook's tube, a board on the wall that was covered with phosphorus started to glow. He asked his wife, Anna Bertha, to place her hand in front of a photographic plate. When he activated the tube he visualized bones and a wedding ring." An invisible wave had somehow passed through the walls of the sealed tube and through a human body. Röntgen didn't invent the sealed tube with the air sucked out that produced the rays. It had been around for a few decades and other researchers had used it and seen the weird glow, but they hadn't discovered its significance. Röntgen rechecked his findings a few times. When he was convinced what he witnessed was real, he announced his discovery to colleagues and newspapers, and X-rays became a phenomenon. Röntgen's revelation led to the development of fluoroscopes and X-rays, though it took a while before we learned of their dangers. When I was a medical student, the head of radiology at Washington U was missing a few fingers. They were damaged when he held people erect while taking X-ray pictures.

By the time I finished medical school, X-ray films were taken from many angles. Air (lung) was black. Bone was white, and tissue gray. By altering the focus of our tube, we could get a sharp view of various depths of a body. As computing became faster and programs became increasingly sophisticated, algorithms were added. Now, physicians can get deep, detailed views of sections of the body, head, and limbs. Modern hospitals use X-ray beams (CAT scans), medical radar (ultrasound machines), and strong magnets (MRIs). The MRI machine uses strong magnets that "constitute a serious, yet invisible danger to people with metal in their body." The machine has to be installed in magnetically shielded rooms, and it's an expensive piece of equipment. It's currently manufactured by at least five major companies, and depending on the size and power of the magnets, the machine alone costs between $250,000 and well over $500,000.

In the last seventy or so years, X-rays have allowed interventional physicians to thread a catheter into a leg artery and advance its tip to the vessels of the brain, heart, and abdomen. A bulge in the main abdominal artery, an aneurysm of the aorta can be replaced surgically, but it can also be splinted by a synthetic expandable stent. The device is inserted through a leg artery and becomes the aorta's inner wall. "In 2003, the interventional approach passed open aortic surgery as the most common technique for repair of abdominal aortic aneurysm, and in 2010, the procedure accounted for 78 percent of the repairs of all U.S. aneurysms that had not already ruptured."

Experts have learned how to thread hollow catheters through one of the four large arteries that supply blood to the brain. Two carotid arteries run on each side of the front of the neck, and two "vertebral" arteries course through openings in the vertebrae of the spine. When they reach the skull, all four vessels communicate with one another.

One in fifty of us will develop an aneurysm, a weakness in the wall of an artery supplying the brain, and 1 to 3 percent of the aneurysms will eventually "split open." Thirty thousand aneurysms rupture each year, and fifteen thousand people die before they reach a hospital. Another fifteen thousand stop leaking but, by one estimate, 7 percent start bleeding again a few days later. Surgically clipping a "ruptured" aneurysm is risky, but it's often necessary. Nowadays, potential ruptures can be prevented by filling the ballooned area with clotted blood. Some interventional radiologists know how to slip small platinum coils into aneurysmal defects and detach the coil from the insertion gadget. When they are properly placed, the coils cause a thrombus to form in the ballooned area. The device that allows radiologists to disengage the wire and leave it curled in the aneurysm was developed by an Italian neurosurgeon named Guido Gugliemi. He was born and trained in Rome and originally planned to be an electronic engineer, but he ended up going to medical school. While there, he was drawn to the brain's wiring, the "millions of relays and millions of wires that transmit electricity and are connected to one another." As a neurosurgeon, he witnessed and treated ruptured aneurysms and found brain surgery to be difficult and bloody. The engineer in him thought there must be a way to use the interventional approach and induce a clot to form. His initial approach involved magnets and it didn't work. He then got financing, moved to Los Angeles with his wife and children, and kept trying to solve the problem. Using his knowledge of soldering and electricity, he spent countless nights and weekends in a lab and ultimately developed a wire that he could detach "electrolytically." Then he moved his family back to Rome. In the U.S., the platinum wires are now commodities that sell for $500 to $3,000, and the cost of their insertion runs $30,000.

A six-year-old seemed cross-eyed and saw an optometrist who realized the child had a sixth-nerve palsy. The nerve that controls the lateral movement of his eye wasn't working and that set off all kinds of alarms. He was immediately sent to an ophthalmologist. The physician detected evidence of brain swelling and ordered an emergency MRI. That night, one of the two hospital pediatric neurosurgeons told the parents their son had a brain-stem tumor. The child was admitted to the hospital and started on high-dose steroids to decrease the pressure inside the skull. A few days later, two pediatric neurosurgeons at Oakland Kaiser Hospital made a small opening in the rear of the child's neck and removed the back of a vertebra. They then sucked out the cells of the medulloblastoma. (It is a friable malignant tumor.) They rolled an MRI machine into the operating room and saw that a tiny trace of tumor had invaded his spinal canal and was still present. The vertebrae was replaced and fastened, and the wound was sewn shut. Postoperatively, the child had obstructive hydrocephalus—fluid was not flowing from one chamber of the brain to the next one. It was possible he would need a shunt that drained the brain fluid and poured it into to the space in the abdomen called the peritoneum. As a first step, under anesthesia a small hole was drilled in his skull. A thin sterile video scope was passed through the skull and shoved into the fluid-filled chamber in the middle of the brain. The scope was then maneuvered through the narrow passages that connected the ventricles, the other fluid-filled brain chambers. Cobweb-like tissue was blocking the flow. It was pushed aside, and the liquid started circulating normally. A shunt wasn't needed and the exhausted, emotionally traumatized parents took a deep breath and renewed their resolve to conquer the disease.

After the child recovered from the surgery, his insurer paid about a quarter of a million dollars for proton beam radiation to destroy the small segment of tumor that remained. It couldn't be surgically resected without risking major neurologic damage, and proton beams, unlike radiation, give off most of their energy in a quick burst to a precisely focused part of the body. The approach decreases the amount of radiation to healthy tissues around the treated area. The child was sent to the Seattle Proton Therapy Center. The facility has ten-foot-thick, lead-lined concrete walls. They isolate a particle accelerator that harnesses and fires protons generated from hydrogen gas. It was one in a national wave of costly commodities funded a decade or so back by private investors and lenders. The Seattle facility treats five hundred people a year. "In 2018, after a net loss of $81 million over the prior two years, its original backers were handed a $135 million loss as part of a negotiated Chapter 11 bankruptcy." Then the facility was able to continue to function. Proton radiation uses the positive particles from hydrogen atoms. Scientists shoot the protons through a ring of circular tubes whose walls are made of magnets. As their speed approaches two-thirds the speed of light, the protons become energized. When they crash into tissue, they destroy it. Unlike X-rays, whose beams harm everything in their path, well-aimed protons damage their target and cause minimal collateral injury. In 1970, a physician named James Slater began working at the medical center in Loma Linda, 68 miles east of Los Angeles. He spent twenty years promoting the concept of a proton accelerator to treat cancer. Eventually, a $100 million facility was built with the help of the national Fermilab outside Chicago and $25 million provided by the U.S. Senate. Some thought that proton radiation would become a moneymaking commodity, and people invested money in various parts of the world. One entrepreneur figured that if a center operated four treatment rooms round the clock they could earn $18 million pretax dollars. The new treatments were twice as costly as standard radiation and insurance companies erected hurdles, but at the beginning of 2020, there were thirty-seven proton therapy centers in the United States and eighty-nine worldwide.

In 1958 Earl Bakken created a battery powered pacemaker that causes the heart to beat regularly. The contraction of the heart is normally initiated by a group of cells in the sino-atrial node. They discharge one or more times a second. The electrical impulse they create flows down the conduction fibers and causes the heart muscle cells to shorten in unison. In 1958, Bakken made his device by modifying a circuit for a metronome, a contraption musicians use to keep the beat. He followed plans he saw in a magazine called *Popular Electronics*.

As a teenager Bakken was the "nerd who took care of the high school public address system, the movie projector, and other electrical equipment." During the Second World War he served in the army signal corps. When the conflict ended he courted a medical technologist. He visited her at the hospital and met doctors who had devices that were broken. He repaired them and realized that hospitals need people to keep their medical equipment working. He set up a shop in his garage and, among other pieces of equipment, made a pacemaker that worked well in animals. At the time Doctor Walt Lillihei at the Mayo Clinic was learning to patch the gap between the ventricles, the two largest chambers of the heart. The doctor discovered that a suture can easily capture the bundle of His. The electrical fiber runs along the border of the defect. If a surgeon ties a knot and constricts the main fiber, the electrical energy can't flow. Hearts stop beating or contract very slowly and children die. In April 1958, Lillihei "was performing open heart surgery and a child developed heart block. The surgeon had seen Bakken's device work on dogs and he told one of the residents to bring in the wire and box." It worked and the pacemaker era was launched. Over the last sixty years pacemakers have become commonplace. There are multiple manufacturers in the U.S., Europe, and Asia, and more than 200,000 devices are planted annually. The devices are used when a lack of effective electrical flow complicates coronary disease, cardiomyopathy and a number of other diseases. Very slow rhythms caused by a malfunctioning sinus node or a conduction problem can cause dizziness, fainting, and exercise intolerance. In the U.S. cardiologists spend a lot of time, inserting the pacemakers through a leg or arm vein and depositing them in the right atrium. The doctors then periodically check the battery, replace it if indicated, and remotely "interrogate" the pacemakers to see how well they were working. Cardiologists also implant devices that can shock the heart of a person who has a history of recurrent ventricular tachycardia or fibrillation. When the heart beats too fast or the heart muscles don't contract in unison, the ventricle doesn't propel blood. The patient will die if his or her heart isn't shocked and restarted within a few minutes. The defibrillators of the 1980s were slipped through a vein and threaded into the heart. In 2012, the FDA approved a defibrillator that can be placed under the skin of the chest wall. Richard Cheney, the former Vice President of the U.S., wore a ventricular defibrillator. He feared terrorists might try to assassinate him by remotely instructing the device to shock him, so he asked his doctor to replace his defibrillator with one that wasn't remotely accessible. When he was in his sixties, Cheney's heart started failing. It didn't propel enough blood with each contraction. Doctors inserted one end of a tube into his left ventricle and the other end into the aorta. Blood was pumped through it by a small electric motor. It helped for a while, and when he was seventy-one years old Cheney received a heart transplant and did well.

For most of us there are defibrillators in airports and many shopping centers. They're not pricey, and they are easy to operate, but many are afraid to use them. U.C. Berkeley put one near a volleyball court and locked the case so no one can easily steal it. That also means no one can use it if someone drops. Hard to figure what they were thinking. In the absence of a defibrillator, rescuers commonly resort to plan B and give CPR. Their efforts push some blood through the system. My hospital required me to retake a class in CPR every two years. The last time I took the course, the teacher said people who develop a lethal rhythm need to be shocked early. If the jolt is delivered within a minute, 90 percent of people survive (presumably without noticeable brain damage). After two minutes 75 percent survive, and at four minutes only 55 percent.

There are defibrillators up and down the halls of Chicago's O'Hare Airport. A survey performed shortly after they were installed found the gadgets had already been used 14 times, and 9 of those shocked came back to life. When someone witnesses a sudden death, confirms the heart isn't beating, and has access to an AED—automated external defibrillator—they merely need to open the case and listen. An automated recorded voice tells them what to do next:

"Attach the electrode pads.

"Don't touch the patient.

"Analyzing. Shock advised.

"Charging. Stand clear.

"Push the flashing button to deliver shock. Stand clear."

Most people who check out one of the brief YouTube videos on AEDs realize how simple their operation has become. If they've seen episodes of TV shows like *ER*, they've witnessed the drill a few times. Which doesn't mean it's not frightening to perform the task on a live person. Most people don't notice the devices that hang on public walls. A survey in a Netherlands train station (per a project manager from Kings County, Washington) found that half the people questioned "couldn't identify an AED," and fewer than half of them would consider using one of the devices. After 40 years, 300,000 people are still dropping and dying each year, and the survival rate to hospital discharge is 8.4 percent.

When I was in medical school, we were taught that most colon cancers originated near the lower end of the bowel and could be detected by a probing finger or by passing a well-lit, metal, foot-long, hollow sigmoido-scope through the anus. Decades later, doctors used snakelike scopes with lenses connected to fiber-optic bundles. (The eyes of a fly are made up of thousands of individual visual receptors. They work together to create an image.) Doctors learned how to guide the instruments around corners, view the entire large intestine, and biopsy or remove growths or tissue as needed. By the year 2000, our endoscopes had an optic chip near the tip and doctors monitored their progress on a TV screen. The devices could be totally immersed in chemicals that killed any virus or bacteria that hadn't been removed with vigorous washing, and sedating drugs controlled the discomfort of the exam.

In 1998, the forty-two-year-old husband of a famous TV personality named Katie Couric died of colon cancer. She became a spokesperson for early detection of cancer by periodically screening the colons of people without symptoms, and colonoscopy became big business. Gastroenter-ologists became skilled, and we could usually pass the scope to the far end of a colon in five to ten minutes. I typically performed five or more procedures in a half day. A similar chip-based scope allowed us to see the stomach and duodenum, treat bleeding vessels, and biopsy worrisome abnormalities.

The small intestine, the nine- to thirty-foot-long stretch of bowel between the stomach and the colon, absorbs food and fluids and is essen-tial to life. It can be partially visualized with a scope, but passing the instrument is difficult and tedious. There's a "capsule" that allows us to view the bowel another way. It's produced by a high-tech company head-quartered in a once sleepy Israeli town in the Jezreel Valley, a village where, in the 1970s, immigrants from Yemen sold pita filled with falafel in the sun-filled town square. The small bowel device and its function are best understood by comparing it to an iPhone. The phone can snap a picture and send it by text or email to another iPhone halfway around the world. The Israeli company uses a camera and similar technology, packaged in a capsule. The "pill" also contains a flash so it can take photos in the dark. There's a "transmitter and batteries that last eight hours." A person with a potential small bowel problem swallows the gadget. As it passes through the stomach and lengthy small intestine, it snaps two photos a second. It can't store the images, so it instantly transmits them wirelessly—like a text photo—to a receiver. It's worn on a belt by the patient who swallowed the pill. At the end of the exam, the capsule is passed in the stool. The storage disc is then plugged into a computer and a program turns the thousands of snapshots into a video that can be viewed and interpreted by a knowledgeable physician or technician.

A medical "device," according to the FDA, is something used to help "diagnose, treat, mitigate, or cure" a human or animal disease or condition. The category includes everything from instruments and apparatuses to machines or implants. Chemicals that don't have to be metabolized to work are devices. Before they are sold in the U.S., "devices" must be registered, listed, and correctly labeled. Production has to be performed "in accordance with good manufacturing practices," and if something goes wrong, if there is an adverse event, the FDA must be notified.

Each year, "over four thousand new, low-risk (class I) devices are marketed." They don't need to be approved, but recalls and problems need to be reported to the agency. At the same time, high-risk devices go through an extensive authorization process and fifty to eighty of them are approved by the FDA annually. Most of the other 3,500 devices that are "permitted" each year are minor modifications of an existing product. In terms of safety and effectiveness, they are "substantially equivalent" to the currently used item. The verification process they go through, called 501k, is not particularly stringent. Only 8 percent need "special controls and clinical data." "Bench testing" is sometimes adequate, but some gadget-tool-appliances have to be checked out "under conditions of clinical usage." When lasers replaced heated wire cautery, they were approved by the 501k process.

The substantially equivalent requirement has periodically been misapplied. On occasion, a manufacturer obtained FDA clearance without revealing the actual use of the device. At other times, the FDA and the manufacturer misjudged. The new product didn't seem that different, but it was, and it should have been extensively tested before it was widely used. In hindsight, that probably was the scenario when the metal-on-metal hip prosthesis was approved. For years, surgeons had replaced painful arthritic hips by inserting a metal stem into the femur and by covering the cup-shaped socket in the pelvis with a "plastic (polyethylene)" insert.

It made sense to many that if both articulating surfaces were metal the joint would last longer and work better. The manufacturer and FDA apparently decided that producing a socket that was metal, not polyethylene, was a minor change and that the modified prosthesis was actually "substantially equivalent." They didn't think it was necessary to test people. Cobalt and chromium alloy prostheses received FDA clearance in July 2008 and were implanted in 100,000 patients. Over time, the metal on the connecting surface of the bone tended to erode. Particles "migrated into the surrounding tissues and the bloodstream. 21 percent of the alloy prostheses had to be replaced or revised within four years and 49 percent within six years."

Transvaginal mesh was also approved by the 501k process. Surgeons used it when they operated on pelvic organs that prolapsed, slipped down, and were protruding into the vagina. No clinical trials were conducted, and between 1985 and 1996 sixty-one products were marketed. The mesh is made of a type of plastic called polypropylene, and over the years it caused a number of "adhesions, bowel obstructions, and infections." In January of 2016 the FDA issued a high-risk warning, and in April 2019 the agency told manufacturers to stop selling the material.

Manufacturers and clinical facilities have to report device-related deaths and serious injuries to the FDA. In 2002, the agency received 2,500 reports from clinical facilities and 3,500 from consumers. "Over one thousand devices are recalled each year." Half have low-risk drawbacks. Most of the rest are intermediate risk. And ten to twenty of the problems are serious. Recalls require manufacturers to halt production and dissemination of the devices, and they are supposed to alert clinicians. Post recall, doctors and nurses are supposed to keep an eye and ear out for difficulties with "heart valves, joint prostheses, implants, cardioverters, defibrillators, respirators, infusion pumps, hemodialysis equipment, cutting and coagulation equipment, endoscopes, etc."

In 2020, a coronavirus that modern man's immune system had not previously encountered jumped from a bat to a man. It often made people very ill and was extremely contagious. At times it caused an infectious process that filled the lungs with fluid and made oxygenation impossible, and many were saved by respirators that mechanically ventilated lungs for days to weeks. In hard-hit areas like Italy and New York, there apparently weren't enough ventilators for everyone who needed them, and the president, the press, and politicians regularly attacked and blamed one another. The nation was caught shorthanded even though government officials had anticipated a potential need for the devices and had ordered a number of them a few years earlier. By April 20, few had arrived and people wanted to know who screwed up. Shortly after the 2003 SARS respiratory epidemic, the Department of Homeland Security decided to stockpile an additional seventy thousand respirators thinking they might be required in a moderate influenza pandemic. Before the government ordered the machines, a panel of experts decided which bells, whistles, and capabilities the respirators would need. The machines on the market were costing $10,000 a unit, and the group thought they were overpriced and decided new respirators should not cost more than $3,000 each. In 2008, the government accepted bids on the project and ultimately signed a contract with a small outfit in Costa Mesa, California, a company that was "small and nimble." After the deal was inked, the government gave the company, Newport, the $6 million they needed to develop the machines. The rest of the money would be paid when the devices were delivered. Reading between the lines, it sounds like the people at Newport knew they would lose money on the deal but they thought they "would be able to make up for any losses by selling the ventilators around the world."

Research started, and every three months, government officials visited Newport's headquarters. "There were monthly scheduled requirements and deliverables." In 2011, three prototypes were sent to Washington and the company planned to "file for market approval and start producing the machines in the fall of 2013." Then the Japanese who owned Newport sold the company to Covidien for a little over $100 million. The new owner asked the government to provide additional funding and a higher sales price, and the U.S. gave Covidien an additional $1.4 million. The following year, Covidien decided the deal "was not sufficiently profitable" and they wanted out of the contract. The government agreed and awarded a new contract for $13.8 million to the giant Dutch company Phillips. In July 2019, the FDA signed off on the new Phillips ventilator, the Trilogy Evo, and the government ordered ten thousand units. They set a delivery date in mid-2020, but in January 2020, a major coronavirus epidemic started spreading out of China. On March 31, 2020, a Pennsylvania subsidiary of Phillips was producing cheap portable ventilators, but the commodities had now become more valuable and the company sold the ventilators to other countries. In March 2020, President Trump invoked the Defense Production Act and told General Motors to make the ventilators.

Devices are responsible for more than 5 percent of U.S. health care spending. In 2016 we spent over $170 billion for medical devices and in-vitro diagnostics and in 2019, the world's top ten device makers had over $209 billion in revenue. In Europe, medical care is usually state run. When countries have options, when two or more companies make acceptable devices, administrators can more effectively use price as a bargaining factor. Not surprisingly, European nations pay less for medical devices than we do in the U.S.

To help make health care affordable to all, the Affordable Care Act (Obamacare) enacted a 2.3 percent medical device tax. It was briefly collected then put on hold, and recently was repealed. The world's largest device maker, Medtronic, the Minnesota Company with revenues of about $29 billion in 2019 spent a decade acquiring and integrating twenty smaller companies. In 2015, they purchased a company headquartered in Ireland called Covidien, became Covidien-Medtronic, and moved their principal executive office overseas. According to The Street, Medtronic's $49.9 billion acquisition of Dublin-based Covidien—the largest tax inversion deal ever—was going to leave shareholders with a big tax bill, while allowing the Minnesota-based company to pay little or no U.S. corporate taxes. According to Robert Willens, tax consultant and professor at Columbia University, "It is not inconceivable that [Medtronic] may not be taxed at all" on its U.S. operations. In 2018, Covidien-Medtronic, now an Irish company, had net earnings of $3.1 billion. Half the money came from sales to physicians and hospitals in the U.S.

CHAPTER EIGHTEEN

Vision

Blind as a bat—better than a poke in the eye—

During the last 150 years, we've learned how to treat or correct most of the conditions that interfere with our ability to see. Trachoma has troubled mankind for thousands of years. It is caused by a chlamydia, a type of bacteria that lives and reproduces inside the cells of the host. The infection causes eyelashes to damage the cornea and (per Hippocrates) it makes eyelids look like cut ripe figs. Antibiotics kill the responsible bacterium, but it tends to recur, and 2.5 million people currently have the "late blinding stage of the disease."

The prevention of river blindness (onchocerciasis) remains a work in progress. Found chiefly in parts of Africa, the condition is caused by a tiny parasite and is spread by black flies that "breed in fast-flowing streams." In the 1990s, The African Programme for Onchocerciasis Control (APOC) treated more than a million people with the anti-parasitic drug Ivermectin, and it made a significant difference. Globally, it is estimated that 18 million people are infected and 270,000 have been blinded by the condition. There's little available data on infants who survive wars, droughts, and suffer from malnutrition. Some develop vitamin A deficiency, and the membrane that covers the inside of their eyelids dry and scar.

The leading causes of blindness in this country are cataracts, glaucoma, macular degeneration, and diabetes.

Cataracts: To create a sharp image, the eye, like the microscope and telescope, uses two transparent lenses. Cataracts are opaque inner lenses. Most cloud up slowly as we age, though cataracts are sometimes seen in children.

I remember the days when people didn't have their inner lenses surgically removed until they were almost blind. After the cataract was taken out, people needed thick glasses and had to grope around to locate their spectacles when they woke. Harold Ridley of England is the father of the implantable lens. The son of a physician, he spent his early doctoring days working on cruise ships. During the Second World War, he spent 18 months in Ghana and later in Burma, and he provided care for former British prisoners of war who had nutritional amblyopia, lazy eye. At some point, he treated members of the RAF whose airplanes were damaged by enemy fire. He noticed that after a cornea, the front lens of the eye, was penetrated by a piece of the plane's windshield, the acrylic plastic did not cause an inflammatory reaction. Years later, he was removing a cataract and one of his students remarked: "It's a pity you can't replace the cataract with a clear lens." That got Ridley thinking. He started crafting implants from the material that was used to make airplane cockpits, and after he removed an opaque lens, he inserted them into the eye. "Sterilization of the plastic was a major problem," and he was afraid to tell anyone. Powerful colleagues were hostile to the idea of putting a foreign body in the eye. There was a learning curve, but Ridley and a pupil perfected the surgical technique and a company in East Sussex (Rayner) manufactured the implant. In 1981, the FDA approved the use of implantable lenses in the U.S., and American eye surgeons adopted the approach. By 2015, nine thousand American ophthalmologists were replacing 3.6 million lenses a year. Worldwide, 20 million cataract surgeries are performed annually. In the U.S., most surgeons numb the eye, insert a small ultrasound probe, and phaco-emulsify (liquefy) the dense lens. (In poorer countries, phaco-emulsification is less common.) Then the doctors suck out the debris, insert a small plastic or silicone lens, and, if necessary, sew the incision shut. My ophthalmologist at Kaiser Oakland told me she doesn't specialize in cataract surgery. The eyes she deals with often have additional problems, so on her surgical half days, she only performs nine operations. Each takes six to fourteen minutes. The complication rate for Canadian surgeons who performed 50 to 250 operations a year was eight in a thousand, and it was one in a thousand for surgeons who replaced a thousand cataracts a year. Most Americans who need cataract surgery are of Medicare age, and the government pays $2,500 per eye. Special lenses can cost an extra $1,500 to $2,500. According to Healthcare Bluebook, a website that tells what insurance companies really pay, the average bill for non-Medicare cataract surgery varies wildly in Northern California. The charge for the procedure (operating room, physician, and anesthetist) can be as little as $1,824. A "fair," middle-of-the-road operation costs $3,422, and, in half of the region's hospitals, the price of surgery exceeds $5,000.

In India, a land with over a billion inhabitants, cataract surgery for the poor took a giant step forward in 1983 when an American public health worker named David Green was hired by Seva, an eye care non-profit in the southern part of the country. Green helped bring volunteer ophthalmologists to the local hospital, and they taught local surgeons how to implant lenses. In the process, they helped change attitudes, and people started coming for surgery before they were totally blind. While he was in India, Green met Govindappa Venkataswamy, an eye surgeon and one of Seva's founders. At age fifty-eight, the Indian physician mortgaged his home, built an eleven-bed hospital, and performed five thousand eye operations a year, 70 percent at no charge. Given the need, he was barely scratching the surface. In the late 1990s, by one estimate, 9.5 million people in India were blind as a result of cataracts and 3.8 million were losing their vision annually. The problem was hard to solve because the cost of implantable lenses, $100 to $150 per eye, was too high for the average Indian. In 1992, Green and others established Aurolab. It's a nonprofit manufacturing plant in southern India that produces inexpensive quality lenses. In 2016, Aurolab fabricated 2.6 million intraocular lenses, 10 percent of all the intraocular lenses that are made in the world. They cost a dollar each to produce and were sold for two to four dollars. The majority of the lenses are now "distributed to NGOs in India and in developing countries," and the company is profitable. In 2003, doctors in India performed 3.9 million surgeries in a single year, and by 2006, cataract surgery in India, Nepal, and Bangladesh was costing $20 and the lens was selling for less than $5.

Glaucoma: The painful red eye and blurred vision that signifies an obstruction of fluid flowing inside the eye is a medical emergency. It's the hallmark of angle closure glaucoma and it doesn't occur very often. The more common condition, open angle glaucoma, is no longer thought of as merely a condition where the pressures inside the eye are too high. The middle of the eye produces a watery aqueous fluid. It flows through the pupil, enters the space in the front of the eye, and exits through the spongy tissue that surrounds the edge of the cornea. In people with glaucoma, fluid is overproduced or doesn't drain normally. The retinal nerve layer thins, people lose peripheral vision, and they can substantially lose much of their ability to see. In the Western world, some ophthalmologists spend a year or more learning how to interpret tests that monitor the problem. They become skilled at performing one of many operations, and in knowing when to intervene. Sophisticated machines allow experts to photograph and follow the appearance of the layers of the retina, the nerve-rich stratum that collects the focused light that our brain turns into images. Gadgets that detect early loss of peripheral vision and that measure the pressure in the eye have entered the digital era. The drugs that control the pressure in the eye include beta blockers and prostaglandin inhibitors. Beta blockers cause the eye to produce less fluid and prostaglandin inhibitors promote drainage. In 2004, after the FDA gave Pfizer the exclusive right to market the prostaglandin inhibitor Xalatan, the company sold $1.23 million worth of the eye drops. Before a generic competitor entered the U.S. market, a month's worth of eye drops was costing $80. According to the World Bank, "Almost half the world's population—3.4 billion people—live on less than $5.50 a day." If one of them develops glaucoma, eye drops aren't an option. Laser surgery, when it's available, can increase the outflow of fluid. If that doesn't work, an older operation called a "trabeculectomy" removes a bit of the mesh network and the fluid pours through and can create "a new drainage path." In 20 percent of the people who undergo surgery, the openings stop working during the first year and 2 percent fail each subsequent year. Worldwide glaucoma is one of the leading causes of visual loss and it has blinded an estimated 4.5 million people in India alone.

AMD (age-related macular degeneration): This is a major cause of vision loss in older people. It's a condition where the macula, the part of the eye that provides sharp, central vision, is damaged or destroyed. The so-called "dry" form of the disease mainly affects white people who are 80 or older, and there is no effective treatment.

The less common "wet" form of the disease is in part due to abnormal fragile blood vessels that grow in the retina, the layer of cells that covers the back of the eye. It is sometimes helped by laser coagulation or photodynamic therapy and is commonly treated with Avastin or Lucentis, antibodies that "block" the growth of the new blood vessels. Avastin was originally approved by the FDA because it stopped the growth of vessels that nourish cancers. After it was available, doctors learned it also halted the progression of wet macular degeneration. Once a medication is FDA approved for one disease, doctors are free to follow the science and use it for other problems. Before using the medication, an ophthalmologist needs to evaluate the patient and discuss the risks of administering the drug. On the appointed day, they have to inject the edge of the eye with a numbing agent and pass a needle into the inner cavity. The chamber is full of a gelatinous material known as the vitreous, and injecting the medicine correctly takes time and skill. A special law limits the amount of money Medicare can pay doctors for their time or skill to 6 percent of the drug's price. The small amount needed to treat an eye costs $50, so Medicare could only pay $3 for each injection. In 2006, Genentech, the company that owned Avastin, helped the doctors when they made a virtually identical drug and charged a lot of money for it. Lucentis, like Avastin, is an antibody that blocks the growth of new blood vessels. It's biologically similar, works as well as Avastin, and is FDA approved for wet macular degeneration. In 2014, Genentech was selling the medication for over $2,000, and doctors who used it in their office were able to charge $180 for the visit and the injection. The law helped create a situation where doctors had two virtually identical commodities made by the same manufacturer, and their price per injection was quite different.

Refractory problem: At some point in most of our lives, we can't see well because our eyes are unable to focus light on the layer of cells at the back of our eyeballs. Some people are born with refractory errors. Others find it increasingly difficult to read small print after they turn 40, and they start using eyeglasses. In the 1950s, people started correcting their vision by placing a thin lens on the surface of their eye, and for a decade, contact lenses were small and had to be removed at night. In 1965, Bausch & Lomb bought the rights to soft contact lenses that were created in the kitchen of a Czech chemist. Once they owned the product's license, the once American, currently Canadian, company started a billion-dollar industry. Otto Wichterle, the man who developed the lens, was a Czech dissident who was jailed by the Nazis in 1942. Sixteen years later, he criticized the country's Communist government and lost his university job. After he was fired, Wichterle worked on the kitchen table of his Prague apartment and used a phonograph motor and an instrument made from a child's building kit (similar to an erector set). He produced four hydrogel contact lenses, and when he put them in his eyes, they were comfortable. In 1962, he patented his invention and produced an additional 5,500 lenses. Ever a protester, Otto was expelled from the nation's chemistry institute in 1970 because he supported Czechoslovakia's attempt to become independent of Russia during the Prague Spring of 1968.

When the Cold War ended, Otto resumed his scientific activities. At one point, he met and learned to trust an American optometrist named Robert Morrison. When he was harassed by patent attorneys, Otto asked Morrison to come to Prague and said, "Robert, I have decided that I must give patent rights to the gel to someone who can use them in the Western Hemisphere and, perhaps, in some other areas as well." He wasn't interested in turning his devices into expensive commodities. He just wanted people to be able to benefit from them. Ultimately, the U.S. National Patent Development Corporation (NPDC) bought the American rights to the lenses from the Czechoslovak government for $330,000. Then they sublicensed the patent to the Bausch & Lomb Corporation, and paid Wichterle less than one-tenth of 1 percent of the money the Czech government received. According to his grandson, he "never regretted not making more money from his invention. He led a comfortable life and he enjoyed traveling and visiting scientists and scholars in many parts of the world." His lenses are currently produced by several companies and they will bring in revenues of $12 billion by 2024.

In 1989, Gholam A. Peyman, an ophthalmologist and inventor, patented Lasik, a laser- and computer-assisted device that allowed doctors to peel back a flap of the outer skin of the front lens of the eye, the cornea. The inner corneal layer could then be altered with the beam of a laser. At the end of the procedure, the flap was replaced and eyes could focus better. The inventor, Dr. Peyman, was born in Shiraz, Iran, and went to medical school in Germany. He's a constant innovator and has held more than one hundred patents. In 2010, it was estimated that 8 million Americans have undergone the Lasik procedure at a cost of about $2,000 per eye.

Finally, no sooner is one problem solved than a new one develops. In a country where the incidence of obesity is increasing as a result of our high caloric diets and diminished activity, more and more individuals are becoming diabetic. Some people with long-standing diabetes develop a number of eye problems and can go blind.

Blood sugars usually have to be elevated for ten to fifteen years before blood vessels on the surface of the retina become permeable and leaky and the amount of oxygen that reaches the cells of the eye decreases. "Left untreated, nearly half of eyes that develop proliferative diabetic retinopathy will have profound vision loss."

CHAPTER NINETEEN

Childbirth

I was in the delivery room for his final performance—the last time an obstetrician named Bill Masters, the doctor who went on to become a world-famous sex specialist, helped a child slide through a woman's birth canal. I don't remember the baby's sex or weight or its mother's glee. But I remember Masters presenting the newborn to its mom with the flourish of a circus maestro, and I recall my fellow med student, whose wife was pregnant, fainting and lying on the floor.

As a junior in medical school, I spent three weeks learning how to deliver babies at Homer G. Phillips, the black public hospital in North St. Louis. I and my fellow student were the only white guys in the facility. We wore blue scrubs, napped in a sleeping room on the second floor, and were periodically awakened by a creaky elevator carrying the almost-mother slowly upwards and a nurse yelling, "Don't push—don't push!" We often barely had enough time to wash our hands, but we were always gowned, gloved, and wore a mask. The delivery area was spotlessly clean—the new mothers and infants were honored guests—and the women didn't have to pay a dime.

In 2010, a vaginal birth was costing between $5,000 and $7,000, and C-sections went for about $10,000. In 2006, the 4.3 million births in this country rang up a bill of $14.8 billion. Before Obamacare became law, pregnancy was commonly classified as a preexisting condition. Medicaid picked up the bill if the woman was sufficiently "low income." But some of the uninsured earned a bit too much. After 2010, expectant women could purchase insurance and they couldn't be charged more because they were pregnant. If they wanted marketplace coverage, they had to "enroll in a health plan during the annual open enrollment period that was set by the employer or the feds." During the seven years after the ACA (Affordable Care Act) became law, 13 million pregnant women "gained access to maternity services." Medicaid expansion played a role. In addition to childbirth, Medicaid covers "contraceptive supplies, sexually transmitted infections, screening for sexual violence, and breast and cervical cancer."

Throughout recorded history, women "midwives" have assisted other members of their sex when they give birth. In the late 1800s and 1900s, physicians, virtually all of whom were men, took over. In 1915, 40 percent of all births were attended by midwives, and twenty years later, close to 90 percent of deliveries were performed by male physicians. In the last decade, medical schools have often graduated more women than men, and most of the young gynecologists are currently female.

In the U.S., 33 percent of children are delivered by C-section, and 9 percent of these women had prior C-sections. The nurse midwife who brought me up to date explained that in her practice about 7 percent of women are delivered by C-sections and only one in four hundred women required an episiotomy, an incision to widen the birth canal. Giving birth vaginally is usually painful, and half of the deliveries that were performed by the midwife I consulted had an epidural, an injection of a derivative of Novocaine into the space outside the lower end of the spinal canal to help control the pain. When physicians deliver a baby, the frequency of regional anesthesia nears 95 percent. C-sections, surgically cutting the abdomen and uterus open and removing the infant, carry the risk of bleeding, infection, and of nicking the bowel and bladder. It also costs more than a vaginal delivery. The approach is easily justified when the mother has active vaginal herpes or the infant would have to come out feet- or bottom first, but that's not why there's a "worldwide explosion." In Mexico City, C-sections are performed for 45 percent of births, and in China the C-section rate was 35 percent in 2014.

The U.S. has the unenviable honor of having the "highest rate of maternal mortality in the industrialized world"—17.8 per 100,000 in

2009. It's especially high for African American women. The global maternal death rate has decreased by 44 percent in the last twenty-five years, but each year four hundred women die per each 100,000 live births in the world's twenty-four poorest countries. In recent decades, most Western countries have cut their death rates in half, but in the U.S., the number of women dying has almost doubled. The care of low- and very-low birthweight infants contributed another $18.1 billion to the birthing price tag. Modern doctors have the incredible ability to keep not-quite-ripe small infants alive, and premature newborns account for half a million of the live births in this country. Some of these kids spend weeks in neonatal intensive care units at a nationwide cost of $26 billion. That turns out to be "about half of all the money hospitals spend on newborns." Approximately 1.7 percent of newborns weighed less than a thousand grams when born, and one-half of 1 percent were under five hundred grams—about a pound. Amazingly, 85 percent of the infants "survived to be discharged from the hospital."

Prenatally, doctors and nurse midwives check pregnant women for diseases like HIV and hepatitis B, infections that are easily transmitted to newborns. When a woman is eighteen to twenty weeks pregnant, screening tests are performed for genetic and developmental problems. In the 1970s, obstetricians started to perform amniocentesis when the risk of a genetic problem was high enough. Using an ultrasound for guidance, they inserted a small needle through the abdominal wall, into the fluid that surrounded the fetus, and they withdrew fluid that they analyzed for Down's syndrome and genetic abnormalities. Amniocentesis is, of course, "invasive," and the procedure induces a miscarriage one-half to 1 percent of the time. The second decade of the 21st century saw the emergence of blood tests that analyze fragments of placental DNA floating in the mother's blood. Fetal DNA and placental DNA are identical. By pregnancy week ten, the level of fetal DNA in a pregnant woman's blood is usually high enough to perform an accurate search for chromosomal abnormalities. When a pregnant woman is near birth, ultrasound exams are performed to check the baby's position and detect problems like placenta previa, a situation where the placenta covers the opening of the cervix and prevents a normal birth.

At $10,000 to $15,000 a try, in vitro fertilization (IVF) and other forms of assisted reproductive technology led to the birth of a million babies between 1987 and 2015. Under ideal conditions, the process is 21 percent successful, per attempt. The fear of malpractice haunts the birthing profession. Childbirth mishaps, mistakes, and bad outcomes still account for close to 10 percent of all malpractice suits, and the amount awarded to injured children can easily be a million dollars or more. It takes an immense amount of money to care for a damaged child for eighty years. Not surprisingly, the malpractice insurance rates for gynecologists are among the highest.

In addition to caring for women during the birthing years, gynecologists have traditionally been the primary care physicians of many otherwise healthy women as they age. Among other things, these physicians pay a lot of attention to the organs of conception. Cancer of the cervix is "worldwide the third most common malignancy in women." It's much less common in this country (11,000 cases a year) because many women have regular "Pap smears." That's the test that was developed by a New York cytologist named Georgios Papanikolaou. He graduated from an Athens medical school in 1904 and decided early on that he wanted to be a researcher. When he was thirty, Georgios and his wife, Andromache, came to the U.S. They didn't speak English and "had little money." She got a job as a button sewer for a department store and he tried to sell rugs. That only "lasted a day," and during his first year in the country, he earned money playing a violin in restaurants. When he got a job in the anatomy department of Cornell Medical College, he hired his wife as his assistant. Using Andromache as a subject, he brushed, stained, and evaluated tissue that the cervix was about to shed, and he noticed that cells that were turning cancerous looked different, "bizarre." He eventually was able to teach doctors to recognize alterations that indicated part of the cervix was almost, but not quite, malignant.

Human papilloma is responsible for over 33,000 cervical and vaginal cancers in the U.S. annually. The virus is sexually transmitted, usually causes no symptoms, and clears within two years. About 14 million Americans are infected annually and 80 million are, at least temporarily, sexually "contagious." In 2014 the FDA approved a two-shot vaccine that usually prevents the disease. It works best when it's given to young women before they are likely to be sexually active. It covers genotypes sixteen and eighteen (responsible worldwide for 70 percent of cervical cancers) It also prevents infection with four additional genotypes that account for 20 percent. Some parents feel that by immunizing their daughters they are saying we assume you will become sexually active, and that's a message they'd rather not send. Harald zur Hausen, The doctor who led the effort to establish the connection between the viral infection and cervical cancer, was a nine-year-old German child, living in a heavily bombed town in the Ruhr when the Second World War ended. A top student, he went to medical school then spent four frustrating years as a researcher at the University of Dusseldorf. While he was there investigators in Philadelphia wrote a letter to his chief and offered a lab position for a German researcher. The department head "threw the letter away "and later mentioned the offer to zur Hausen. The young man "fished the letter out of the trash can," accepted, and moved his family to Philadelphia. He later returned to Germany and was instrumental in establishing the relationship between cancer and the viral infection. In October 2020, an eleven-year study in Sweden showed that the human papilloma virus vaccine decreased the risk of developing cervical cancer by 90 percent.

Early on, the tools of the GYN trade relied on feel and a speculum. The uterus and ovaries were felt by trained fingers in the vaginal canal that pushed up and equally aware fingers on the abdomen that pushed down. When the exam was painful or the woman was large or tense, the exam had limited value. Shadows of the uterus and ovaries are now sometimes visualized using an abdominal ultrasound, a CAT scan, or by placing an ultrasound probe into the vagina and watching a TV screen. The charge for an examination with the probe in one location chosen randomly on the Internet runs $200. I don't know if insurance companies will pay for the test in the absence of a clear indication. It has not, best I can tell, become "routine," though actress Fran Drescher and others think it should be. The $6.5 million bill President Bush signed in 2007 "authorized the development of a national gynecologic cancer awareness campaign" but did not mandate screening vaginal ultrasounds. Gynecologists have long evaluated the inside walls of the uterus with an operation known as a D and C. They dilate or stretch the cervical area. Then a sharp instrument is placed inside the uterus and the lining cells are scraped off, collected, and examined under a microscope. The main indication for the operation is unexplained uterine bleeding, which could be caused by cancer of the inner lining wall of the uterus. Nowadays, there's a thin, narrow scope that doctors can slip into the uterine cavity and look for abnormalities. In this country, hysteroscopy is usually performed in anesthetized patients. In Australia and elsewhere, it's sometimes performed with light sedation and numbing agents.

Finally, the gynecologists were pioneers in the use of a tiny incision and a laparoscope (see surgery) to evaluate ovaries, treat cysts, or tie fallopian tubes so a woman could avoid pregnancy. Gynecologic surgery is a relatively large ticket item. In this country, 600,000 women have hysterectomies annually for benign conditions. One hundred eighty thousand (30 percent) of the operations are done for "fibroids"—benign growths that can cause symptoms. Some are performed in an attempt to reduce or eliminate lower abdominal pain, discomfort that is sometimes caused by endometriosis, a condition where the kind of tissue that normally lines the inner wall of the uterus is growing elsewhere and is sensitive to female hormones.

CHAPTER TWENTY

Care of Kidney Failure Becomes a Right

"Wherever the art of medicine is loved there is also a love of humanity."
—HIPPOCRATES

O n September 30, 1972, ten years after I graduated medical school, Congress added an amendment to a Medicare bill and the federal government started paying for the cost of keeping people with kidney failure alive. When I was studying to be a doctor, people with advanced chronic kidney disease died. Doctors only dialyzed people whose renal failure was acute and potentially reversible. If a kidney stopped working and didn't start within a few weeks, dialysis was stopped.

In 1960, a group led by Dr. Belding Scribner, a man who called himself Scrib, made chronic kidney dialysis possible. He used a Teflon catheter as a pipe that joined an artery to a nearby vein. His original device consisted of two lengths of plastic tubing—two cannulas. One length was inserted into a forearm artery and the other into a nearby vein. Both tubes exited the skin that covered the under surface of the forearm. When the tubes were not being used for dialysis, they were attached to opposite ends of a Teflon tube. Blood constantly flowed from the high-pressure artery through the shunt and into the vein. When it was time to dialyze, the cannulas were uncoupled from the Teflon tube and were attached to the dialyzing machine. Then blood flowed from the artery, through the machine, and into the vein. Scrib, the man who invented the shunt, didn't see well. When he was young, his vision was saved by transplants of his cornea, the outer lens of his eyes. After he graduated from medical school, the lenses "became very scarred." He moved to Seattle and lived on a lake near the VA hospital where he was the director of medical research. Each day, he rowed to work.

In 1963, "the Veterans Administration (VA) established thirty dialysis treatment units in their hospitals." Over the subsequent eight years, "Programs sprang up across the country." By 1972, the year Medicare took charge, ten thousand Americans with renal failure were being dialyzed. Medicare spent $28 billion on dialysis in 2016. There are 4,000-plus freestanding centers in hundreds of cities. More than half are owned by Denver-based DaVita, a Fortune 500 company. The other big player, Fresenius, is a subsidiary of a German company that operates centers in twenty-eight countries and also sells the machines and other supplies. Nowadays, before hemodialysis is initiated, an artery and a vein are sewed together under the skin and a large pulsatile blood vessel is created. Patients come to a center three times a week and sit in a lounge chair while they are being dialyzed. Some people opt for peritoneal dialysis and use the lining wall of the abdomen as the filter. Before they start, a physician places a permanent catheter deep in the cavity. Then, usually nightly in the person's home, the patient pours a sterile solution (dialysate) through the catheter and into the peritoneal cavity. Over several hours, the membrane's pores allow waste products like urea and creatinine to seep into the fluid. Then the fluid is drained. In recent years, peritoneal dialysis has been costing Medicare about $53,000 a person per year. People with very diseased kidneys are usually anemic. To stimulate the production of red cells, people on dialysis are given injections of erythropoietin. It's a costly hormone that is normally produced by healthy kidneys.

A fourth of Americans who are being dialyzed die annually, and 80 percent live for less than five years after they start. In France and Japan, the annual death rate is only 7 to 8 percent. On occasion, dialysis is started when people are in nursing homes, and they usually don't do well. In one study, 58 percent died the first year after dialysis was started and 29 percent were less functional.

Nearly 20 percent of people with end-stage renal disease stop dialysis prior to death. That's what happened to the famous humorist Art Buchwald. In his later years, he had diabetes and high blood pressure. When he was seventy-four, he had a significant stroke. Six years later, a leg was amputated. I don't know when he started dialyzing his blood, but in 2006 he was eighty and decided to stop. He apparently described his decision as his last hurrah. Contemplating death, he wrote, "the question isn't where you are going, it's what you are doing here in the first place." Known in the hospice as the man who wouldn't die, he once quipped: "To land a big obituary in the *New York Times* you have to make sure no head of state or Nobel Prize winner dies on the same day," and "Whether it's the best of times or the worst of times it's the only time we've got. The best things in life aren't things." A year after he stopped dialysis his heart stopped beating. The cause of death was kidney failure.

Medicare also pays for kidney transplantation. By 2016, over 13,000 kidneys had been transplanted in the U.S.—5,600 of them came from living donors.

We were visiting an acquaintance who was in the hospital recovering from a transplant. "So," she said, "things kept going wrong with my shunt and my dialysis and it seemed like I was always in the hospital or on the phone complaining. One day I was in the midst of a tirade and the friend I was talking to stopped me cold. 'OK, OK,' she said. 'I'll give you a kidney,' and here I am."

Forty-one years after the Medicare dialysis benefit was initiated, the NIH reported that 63.7 percent of people with advanced renal disease were receiving hemodialysis, 6.8 percent were being treated with peritoneal dialysis, and 29.2 percent had a functioning kidney transplant. Nine of ten transplanted kidneys were working a year after they were implanted, and half lasted at least ten years.

A few years back, I read that about six thousand undocumented Americans had end-stage renal disease. One hundred eighty of them lived semi-near the Texas Baylor emergency room. Each day, many came to the ER and waited. Their blood was drawn, and EKGs were obtained and assessed. The hospital had twelve dialysis chairs, and the people who occupied hospital beds were dialyzed first. Then the sickest of the undocumented were served.

Texas and most other states only detoxify the blood of people whose serum potassium is quite high or whose buildup of toxins is so extreme that death is imminent. Some have to wait until dialysis has literally become a matter of life and death. For context, more than 500,000 undocumented Americans are dialyzed on a regular basis, and almost 11 million unauthorized immigrants live in this country. One in 1,833 need dialysis. Federal funds can't be used to treat them, but California and Massachusetts draw on county taxes or state-allocated Medicaid funds to pay for their care.

CHAPTER TWENTY-ONE

HIV and the Plight of 23 Million Africans

One morning in the 1980s, I attended a conference at the university and learned that a rare tumor, Kaposi's sarcoma, was affecting members of the gay community. Until then, the cancer had chiefly affected old Italian men. Young people were now developing Pneumocystis pneumonia. The microbe responsible for the infection had previously lived quietly in most bodies for a millennium. Problems swallowing were caused by white splotches of yeast that covered the walls of some esophagi. Infections of this sort had occasionally been seen in people who had kidney transplants and were taking drugs that suppressed their immune system, but why healthy young people?

In 1983, scientists in France and in the U.S. at almost the same time isolated the responsible virus. The microbe was invading, controlling, and ultimately destroying T lymphocytes, the immune cells that restrain many of the organisms that live in our bodies. When the T cells are gone our immune system loses its ability to control the microbes and many attack at the same time. In its advanced state the infection is called AIDS, acquired immunodeficiency syndrome.

We learned of cases of the disease in thirty-three countries. Actor Rock Hudson died with AIDS and the Hollywood community rallied.

Our blood supply was tainted. Ryan White, a kid with hemophilia who had received multiple transfusions, developed AIDS and was not allowed to attend classes. A journalist who worked for the *San Francisco Chronicle* published a book called *And the Band Played On*. It called the first five years of the AIDS epidemic "a drama of national failure played out against a background of needless death." Addicts who injected drugs and shared syringes were also sharing the virus, and needle exchange programs were initiated. Condoms were encouraged. Gay bathhouses were attacked by police. Physicians were not allowed to test people for HIV without permission.

The HIV virus, we later learned, is not highly contagious. You could shake hands or hug someone who had the virus growing in their body. Two principles of medicine, however, were turned on their head: In otherwise healthy people, acute infectious illnesses were usually caused by a single microscopic organism. When a pneumonia was cured, a person could stop taking antibiotics. In people with AIDS, antibiotics often suppressed but failed to eliminate the creatures responsible for a disease. The infected individuals were at risk for a recurrence and needed antibiotics for life. Also, when we identified an organism that caused an infection (like Pneumocystis), we weren't out of the woods. The creature had attacked because the immune system wasn't effective. Additional problems were brewing.

Within a few years, medical detectives in Cameroon found chimp feces that contained simian immunodeficiency virus (SIV) whose DNA was identical to the DNA of the most common strain of human immunodeficiency virus (HIV). Many experts believed the source of the epidemic was a chimpanzee that was infected with the simian immunodeficiency virus. In parts of Africa where the creatures were a source of protein, "bush meat," the virus presumably jumped to man when a human was injured by a monkey he or she was killing. Blood containing the simian virus entered the hunter's body, and the virus survived and thrived. Later, the pathogen was sexually passed to one person after another (as explained in the Craig Timberg and Daniel Halperin book *Tinderbox*). Sometime in the early decades of the century, an infected individual moved to Leopoldville (now Kinshasa). In 1920, the town was the capitol of the Belgian Congo and was full of migrants. Scientists figure that by the time the colony became an independent country (1960), an estimated 1,000 to 2,000 people were living with HIV. A physician of the day probably encountered many sick souls who had developed diarrhea, fever, and wasting. There's no reason to believe that anyone at the time suspected their symptoms were the result of a new virus. When the Congo became an independent country, volunteers from Haiti flew in to provide medical care and assistance. One of them presumably caught HIV. He or she eventually went home, and spread the virus in Haiti. A few years later, an unknowing carrier visited the U.S. and Europe and passed it on.

After they identified the likely cause of the disease, scientists learned how the virus methodically infects a cell. HIV is an RNA virus. It uses a special enzyme it brought with it (*reverse transcriptase*) to create a DNA version of itself. The new DNA becomes part of the DNA in the nucleus. It in essence becomes a gene that is passed to a cell's offspring. The new DNA then directs cells to keep making HIV viruses. The viral DNA becomes part of the T cell's nuclear DNA. Like a gene it directs the cell to keep making HIV viruses, and it is passed to the T cell's offspring. Our alphabet has twenty-six letters, and we use them to make words. The DNA in the nucleus of cells uses a four-nucleotide alphabet. In each cell, there are twenty thousand chains of these nucleotides, which are genes. They only occupy up to 3 percent of a cell's DNA, but they are the blueprint, the instructions, that tell the cubicles what to make and do.

R., a gay thirty-five-year-old carpenter who was married to a gay woman and had a daughter, was renting an apartment I owned in the San Francisco Bay Area. I came by to check on something one afternoon and heard coughing from the house. R. let me in and told me he didn't feel well and was having trouble breathing. His condition worried me and I suggested he visit a hospital. He went and the doctors found he had Pneumocystis pneumonia and advanced AIDS. Treatment was started, but he got worse and needed to be intubated. A respirator kept him alive for a week and R. recovered. He was able to leave the hospital, but at the time everyone knew AIDS was a death sentence. After he came home, R. moved out of the apartment, bought a house, fixed it up for his family, and invited my wife and me over for Sunday afternoon tea. He told us that he had long wanted to have a home of his own before he died. He looked good, seemed relaxed, and was proud of what he had recently accomplished. We never saw him again.

The counterattack against HIV started when scientists at Burroughs Wellcome synthesized compounds that hindered the activity of the *reverse transcriptase*, the enzyme that allowed the RNA virus to make a DNA copy of itself. In 1985, they sent eleven promising compounds to researchers at the National Cancer Institute and scientists identified a chemical that worked in the test tube. The medication, AZT, was given to people with HIV, and their lives were prolonged. Twenty-five months later, the FDA approved the drug, and it was marketed by GlaxoSmithKline. The company sold 225 million dollars' worth in 1989.

In the late 1980s and in the 1990s, manufacturers started cranking out (and selling) anti-HIV drugs. Some of the agents targeted protease, an enzyme that plays a role in the production of more viruses. The first protease inhibitors became available in 1996. Raymond Schinazi, an Emory University professor, and chemistry professor Dennis Liotta developed emtricitabine and lamivudine. The drugs that were defective nucleotides and they prevented invading HIV from turning its RNA into DNA. As Schinazi explained, "Everyone was intrigued but skeptical about our work. No one realized the importance of what we had found." Schinazi pushed Emory University to file patent applications. That's what universities were supposed to do. The drugs were a commodity and the law encouraged owners to reap a profit. When the university licensed the drugs to a pharmaceutical company, Emory was paid $540 million. In 1996, a three-drug regimen was shown to successfully suppress the HIV virus, and in advanced nations the disease could be controlled. But the companies that owned each drug's patent charged what they thought they could get away with and the commodities were too pricey for most people in the developing world.

As Nobel Prize winner Joseph Stiglitz explained, patents are created for each nation's needs. They give the inventor a monopoly for a number of years. When they are appropriately designed, they promote innovation and societal well-being. When they are not appropriately designed, people die and innovation is suppressed. The patents in question benefited first-world pharmaceutical manufacturers. Back then, most nations allowed people and companies to patent unique, non-obvious inventions and new medications. India had a different approach. Their patent law was enacted in 1966 when Nehru's daughter Indira Gandhi was prime minister. At one point, she met with the head of the Indian drug maker Cipla. He convinced her to allow inventors to patent the process—the way the drug was manufactured, but not the drug itself. That became and remained the law in India until the country became a member of the WTO—the World Trade Organization. If the country wanted to be able to trade with the U.S. and Europe, they had to sign on. After India joined the international group, the country was forced to change their patent law so drugs themselves could be patented.

In September 2000, years before the rules changed, Yusuf Hamied, the CEO of Cipla, was invited to the European Commission in Brussels. He participated in high-level talks with health ministers and heads of large pharmaceutical companies. Cipla had been founded in 1935 by Hamied's father, and it was a major pharmaceutical company. The meeting was supposed to discuss access to medicines in the developing world. After the leaders of various gathered companies made their remarks, Hamied spoke, saying he represented the developing world and an opportunity. He offered to provide the three anti-retroviral drugs that suppressed HIV at a cost of $800 a year and was willing to supply needed medicines to pregnant women so they would not infect their unborn child. Poor nations could be taught how to create pharmaceutical factories. The proposal was serious and significant, but it was ignored. In the early 2000s, according to Denis Broun, M.D., of Unitaid, the powers that be believed "treatment for AIDS was something for the rich. It was unthinkable for Africans." Yusuf Hamied believed Africa was being taken for a ride. A year after the Brussels meeting, James Love, an AIDS activist, called Hamied and asked how cheaply he could produce the three drugs that suppressed the virus. The key to pricing in medicine, Hamied told him, is the cost of the active pharmaceutical ingredients. If you can get them cheaply, the end product is cheap. Hamied offered to have his company, Cipla, pay the cost of manufacturing a generic regimen. He would only charge for the material. Nevirapine would cost $0.65 a day; 3TC, lamivudine, $0.35. And there would be no charge for d4T, stavudine. The materials for the drug were too cheap. In other words, the three-drug regimen would cost $350 a year. Donald McNeill of the *New York Times* felt the offer "was a watershed event," and he put the price of generics on the front page of the *Times*. Papers around the world spread the news.

Shortly thereafter, Peter Mugyenyi, a Uganda physician and director of the continent's largest research and treatment center, decided to take matters into his own hands. "I knew where drugs were, and as a doctor it was my job to save my patients' lives." He contacted Cipla in India, and in defiance of patent laws ordered the medications. When they arrived at the country's airport, they were impounded and the doctor who came to pick them up was arrested. He refused to leave the airport without his drugs and the authorities eventually relented. He distributed the medications, and other nations joined Uganda. To many, it seemed like the blockade for inexpensive drugs in Africa was broken. In 2002, Kofi Annan, the diplomat from Ghana who was the secretary-general of the United Nations, proposed a Global Fund to buy the medications. The U.S. insisted that the fund could only buy branded medications or they would pull out. In 2003, South Africa's competition commission ruled that GlaxoSmithKline and another company had violated the country's anti-competitive act. Glaxo was charging excessively high prices and was refusing to license their patents to generic manufacturers in return for reasonable royalties. The company eventually agreed to allow three generic manufacturers to make and sell three of its AIDS drugs, and the company took a 5 percent fee.

Prior to 2003, the U.S. hung tough. Then the Irish singer Bono got together with one of the day's more influential Republican senators, Jesse Helms, and attitudes changed. When they met, the senator was 80 and walked with a four-pronged cane. He was a right-wing evangelical Christian who had exploited racial prejudices in his election campaigns and had called homosexuals "weak, morally sick wretches." Bono, by contrast, had publicly supported Greenpeace, Amnesty International, and had joined Jubilee 2000, a forty-country movement that advocated canceling third-world debt for the millennium. At one point, the Jubilee campaign asked Bono to get the Baptist Nigerian president to write a letter to Baptist churches across southern U.S. states. He was supposed to explain the biblical principles behind debt cancellation. The Baptist leaders listened, and Bono suddenly had access to a lot of strongly Christian Republicans. That's why he was able to meet and speak with Jesse Helms. Helms had been very tough on the concept of foreign HIV drug assistance. "He's a religious man," Bono said, "so I told him that 2,103 verses of scripture pertain to the poor, and Jesus speaks of judgment only once. It's not about being gay or sexual morality, but about poverty. I quoted that verse of Matthew chapter 25: 'I was naked and you clothed me,' and he was in tears. He later publicly acknowledged that he was ashamed."

After the meeting Vice President "Dick Cheney walked into the Oval office, and told President Bush that, 'Jesse Helms wants us to listen to Bono's idea."That led to negotiations and Bush's 2003 plan. That January, in his State of the Union message, President Bush announced his policy toward HIV had changed. His words: "Today on the continent of Africa, nearly 30 million people have the AIDS virus, and across that continent only fifty thousand are receiving the medicine they need. Many hospitals tell people: 'You have AIDS. Go home and die.' In an age of miraculous medicines, no person should have to hear those words." "Anti-retroviral drugs can extend life for many years. And the cost of those drugs has dropped from $12,000 a year to under $300 a year. Seldom has history offered a greater opportunity to do so much for so many."

Bush would ask Congress to spend $15 billion over five years to combat the disease. Since its creation in 2003, the "President's Emergency Plan for AIDS Relief (PEPFAR)" has received more than $70 billion in congressional funds, and in fiscal 2017, the nation contributed $6.56 billion. Bush's embrace of the under $300 number caught the drug industry by surprise. They fought back and within days the administration had changed its approach. Rather than generic anti-retroviral drugs, the U.S. government money would be used to buy high-priced branded medications. As Bill Clinton later put it, "If you ran the numbers, there was no way the money was enough to save the number who had to be saved in a hurry . . . it would never be enough unless they bought generics." Clinton was not president and was not seeking office, and he decided to ignore the politics of the situation. In late November 2006, he announced an agreement between his foundation; two Indian drug makers, Cipla and Ranbaxy; Aspen Pharmacare of South Africa; and Matrix Laboratories of Dubai. The companies agreed to make pills for children that combined three HIV drugs and cost $60 a year. Two million children in Africa had been infected by their mothers and only 10 percent were receiving drugs. Without treatment, 80 percent would be dead by age five.

The companies also apparently agreed that, "The cost of anti-retroviral drugs (in general) was going to drop to $140 a year, and pills would cost $0.36 to $0.38 a day." The cost of constructing the factories and producing the medicines was paid for by a $35 million grant from an international drug-buying consortium and $15 million from the Clinton Foundation. The funds guaranteed the volume of drugs purchased would be "high enough to justify the lower prices." "Orders for large quantities of generic drugs were critical to bringing African prices for anti-retroviral treatment below $100 per person per year." "After that the cost difference between branded and generic anti-retrovirals, and the scale of human tragedy in Africa made it impossible for donor funds to spend vast sums of money on expensive drugs. The global fund and PEPFAR eventually committed themselves to buying generics, and the number of people treated exploded."

In 2017, per the U.N., 19.5 million people, more than half those infected, were being treated, but 2 million additional people had acquired the disease. More than 25 million HIV carriers lived in sub-Sahara Africa. That year, worldwide, 75 percent of people who carried the virus knew they were infected. The U.N. thinks they can end the epidemic if 90 percent of those infected know they are infected. Then 90 percent of those infected need to take anti-retroviral drugs that effectively lower the measurable blood level of HIV 90 percent of the time. Seven countries have achieved the 90/90/90 goal. The CDC (Centers for Disease Control) thinks that 1.1 million Americans are currently infected, and 40,000 inhabitants will acquire HIV each year. Between 83 and 88 percent of those infected have been diagnosed, and 85 percent of the people regularly take medicine that controls the virus, but only half take enough pills in the right dose. The drugs that turn HIV into a chronic controllable condition are commodities, products, and people with "decent" health insurance are often required to pay two-thirds of their cost. That comes to about $20,000 a year, or $360,000 in a lifetime. In the appropriate age and income situations, Medicare and Medicaid supply the medications and the nonprofit Ryan White Foundation helps people who can't afford the drugs. I don't think most insurance pays for the drugs that prevent HIV. Gilead marketed a combination of two antivirals that significantly reduced the risk of sexually acquired HIV-1 infections. In 2019, the medications had a list price of $24,000 a year and were bringing in annual revenues of close to $3 billion. There were assistance programs for the poor. I'm not sure how hard it was to get the drugs for free, but when an $8-a-month generic form of the drugs was introduced in Australia in 2019, the number of new cases of HIV "dramatically declined." (A vaccine that prevents HIV was never developed. See chapter on vaccines for details.)

CHAPTER TWENTY-TWO

The Right to Emergency Care

I n 1986, a well-intentioned but unfunded congressional bill called the Emergency Medical Treatment and Active Labor Act (EMTALA) gave "everyone" the right to receive emergency treatment.

"Everyone" includes the insured and uninsured. Citizens, foreigners, and "illegals." People who appear in the emergency room and those who called an ambulance. It applies to all hospitals that take Medicare funds—in other words, almost every hospital in the country. The care must continue until the person is "stable."

Before the law was passed, some hospitals wouldn't deal with people who couldn't pay. Sick, unstable, uninsured people were "dumped." People were "transferred from private to public hospitals for financial reasons without consideration of their medical condition or stability." Some of them died.

After the law was passed, hospitals were required to medically screen any person who came to their emergency room and asked to be checked out. If the person had an urgent problem, the hospital had to stabilize the situation. A person with a ruptured appendix had to be treated medically and/or surgically before he or she was discharged. Someone suffering an acute myocardial infarction (if medically appropriate) had to be catheterized, stented, and observed before he or she was released. If the hospital didn't have the expertise or equipment to care for a person, they were required to transfer the patient to an appropriate facility. Hospitals were in violation of the law if they delayed services "in order to inquire about the individual's method of payment or insurance status."

The sick person didn't have to be in the emergency room to qualify for care. They could be in almost "any area of the hospital campus," including structures and areas that were not strictly contiguous with the main building. They just had to be within 250 yards of the ER or of the hospital-owned and -operated ambulances.

On-call doctors were often on the hook. In some states, they had to show up within 30 minutes of being notified. There are rules, and physicians face fines if they violate them. We're talking up to $50,000, money that is not covered by malpractice insurance. Transfers are possible with a patient's consent, if the person is "stable."

After the patient is discharged, the hospital can send a bill—and they do.

A friend awoke one night with chest pain. He was sweating and dizzy, and his daughter called 911. The ambulance arrived quickly. The paramedic determined he was having an acute myocardial infarction, and he was rushed to the county hospital where a skilled cardiologist was ready to act. Within minutes, a tube was inserted into an artery in his leg and threaded up the aorta. A balloon on the catheter opened the artery and the stent kept it open. Blood flowed to the oxygen-starved portion of the heart muscle, and it looked like the heart had escaped serious injury. He was observed for a day, went home, and was sent a bill. My friend lived from paycheck to paycheck, didn't own a home, and had allowed his insurance to lapse.

U.S. hospitals and doctors accept the amount reimbursed by Medicare and Medicaid as payment in full. (That's the law.) There's also a website that tells how much insurance companies really pay. Called Healthcare Bluebook, it allows people to punch in their medical procedure and zip code. The program provides a dollar amount hospitals receive for many operations and procedures. If the website is correct, the bill my friend with the heart attack received was outrageous.

The "fair price" for a coronary angioplasty and three days in a hospital in his zip code (per the Bluebook) was $24,853. (That covers the cost of the catheterization room, the equipment, and the hospital bed.)

The "fair" physician remuneration for stenting a single vessel was $1,105. (Add 25 percent for additional arteries.) The "fair" anesthesiologist fee for close to three hours of work was $1,325.

My friend's bill for his heart stent and two days in the hospital was in excess of $100,000. As I read the Affordable Care Act, the bill appears to be unlawful, but I'm not planning to get into the weeds of the situation.

What will it cost if your appendix bursts or you have a severe asthma attack when you're in Paris? Not much. Health care is good in France. An emergency room visit costs about $120, and a doctor's visit goes for less than $30.

When someone has a medical emergency they often go to the nearest hospital. When they don't have health insurance they get a bill that is often enormous. When people do have health insurance twenty percent of the time they land and are cared for in a facility that is not in their insurance company's network. The amount charged for their care can exceed the amount their health plan is willing to pay and the ill or injured person receives a "surprise" bill. Insurance companies have long argued that "networks" allow them to control the cost of care. They meet with hospitals, agree on fees, and sign in-network agreements. New York and a number of other states have laws "that cap or limit charges for services that are delivered out-of-network, especially for emergency care."

California law, AB 1611, would have limited the fees hospitals charge when someone has insurance but ends up in an emergency room that's "out-of-network." The law said hospitals and doctors can't bill for more than the amount they receive from in-network cost sharing. Shortly before the legislation was going to be presented to the state Senate health committee, the sponsoring legislator "pulled" it. He said there was "insurmountable" opposition from lobbyists and CEOS of California hospitals." They objected to a provision that limited the charges for out-of-network emergency services.

In 2020 there were more than 4,500 emergency departments in the country, and they were staffed by about forty thousand physicians. Many hospitals "contract out" their emergency providers." Their services are commodities, and two large companies EmCare and TeamHealth control 30 percent of the physician market, establish fees, and send bills.

"Between 2010 and 2016 out-of-network bills rose more than ten percent and the number of inpatients who got an out of network bill for some of their care rose from 26 to 42 percent. During the six year period the cost of an ER visit increased, on average from $220 to $628 and hospitalization charges rose from $800 to $2,000."

As part of a non-emergency medical encounter, doctors or services at an in-network facility are often provided by out-of-network providers. It's claimed that 9 percent of the time the insurer only pays part of the charges, and the person who has insurance gets an additional bill.

A fifty-year-old woman in California told me about her recent checkup by her physician. While there she had a Pap smear, a screening test for cancer of her cervix. The exam was performed by her physician but it was sent to an out-of-network lab and it led to an inflated bill.

California bill AB72 doesn't allow the out-of- network doctor to charge more than an in-network physician could charge. The law is supposed to protect patients who get their care at an in-network facility, and one of the doctors is out of network. It's a state law. As such it doesn't apply to "self-insured, employer-based policies." A lot of people in this country are covered by those kinds of policies.

In late 2019, two U.S. congressional committees were unable to put forward legislation that would have based the rate sick people are billed on the amount doctors and hospitals usually charge. Their efforts were thwarted by private equity investors who own practices, free-standing emergency rooms, and air ambulances. These entrepreneurs advertise and contribute to election campaigns. They argue that in the land of free enterprise, providers should decide how much they will charge. When people disagree they should arbitrate.

A sixty-year-old grandmother who had been married for twenty-six years developed abdominal cramps and nausea. At the time she was retired and was touring the Southwest in an RV with her husband. She visited an in-network gynecologist in the nearby city of Carlsbad New Mexico. As part of the workup the doctor sent a vaginal swab to a lab. It was checked for six sexually transmitted diseases, and all the tests were negative. The woman didn't understand why the doctor requested the tests—but whatever.

The initial bill for her exam and tests exceeded $12,000. Her insurer Anthem paid over $4,000 on a "negotiated rate of over $7,000." But the woman still owed over $3000 for the lab tests. The amount charged was clearly excessive. Medicare and Medicaid usually pay $40 for each of the six lab tests the doctor ordered. When the woman protested the clinic threatened to send her bill to a collection agency. I heard her story on NPR. It was used as an example of how far some clinics will go when they are allowed to charge what they think they can get away with. To get the details check out the web account: "Retiree Living the RV Dream, Fights a Nightmare $12,387 Lab Fee. December 23, 2020."

The outlier—the hospital that started billing fairly: In early 2019, San Francisco General Hospital, the city's trauma center, was under pressure for its unfair billing system. The hospital had no contracts with insurance companies. All the care it provided was out of network. As a result, the city could bill as much as their self-determined rates allowed. Health Insurance companies would then kick in as little as their self-determined policies permitted. The difference, thousands of charged and unpaid dollars, became the burden of the ill and injured people who showed up in the facility's emergency room.

"Three simple appendectomies left stunned patients on the hook for bills ranging from $54,000 to $92,000." Well-meaning doctors, nurses, clerks, and ambulance drivers were treating people's physical maladies and destroying their economic well-being.

The city owned the hospital, and taxes paid for some of the care. In 2019, the hospital leaders made a decision that was unique and unprecedented. In place of being fiscally "prudent," the administration stopped billing poor people. They charged a maximum of $4,800 to people who earned less than $121,000 (single person) or a family earning less than $250,000.

As the local newspaper, the *Chronicle*, put it, injured parties "can (now) rest easy. If you're hit by a car, shot, fall off a roof or suffer any other major injury, you can now receive top-notch medical care at the city's only trauma center without risking bankruptcy."

The effect on the hospital's income was a fifth of a percent. I don't know of other hospitals that are following their lead. It's inconsistent with the idea that emergency rooms and hospital care are supposed to be moneymaking enterprises.

The rules of the game are apparently on the verge of changing. In December 2020, U.S. congressional legislators managed to add language that eliminates surprise medical bills to a $900 billion COVID relief bill. President Trump eventually signed the law. "Instead of charging patients, health providers will now have to work with insurers to settle on a fair price." The law will won't take effect until 2022. It will apply to doctors, hospitals, and "air ambulances, though not ground ambulances."

CHAPTER TWENTY-THREE

Hospitals

C harity Hospital, the country's second largest, provided medical care to the poor and uninsured of New Orleans and had one of the nation's busiest emergency rooms. Rebuilt for the sixth time in 1939, it had 2,680 beds, and was a major teaching hospital for Louisiana State University. In August 2005, Hurricane Katrina struck the city and flooded the first floor of the structure. Aside from a few generators for breathing machines, there was no power, no lights, and food was scarce. On more than one occasion, the thermometer topped 100 degrees and toilets didn't flush. There was no water for hand washing, and the porta potty at the end of the hall smelled. One doctor wrote, "Our 250 patients were evacuated by twos or threes in boats, and it took nearly a week before the final 200 were rescued." At the same time, helicopters were efficiently evacuating patients from the roof of the nearby Tulane Hospital, a facility that was 80 percent owned by the for-profit Hospital Corporation of America.

Almost a fourth of the hospitals in the U.S. are stock-payer owned. HCA, Hospital Corporation of America, is the nation's largest for-profit collection of hospitals, surgery centers, freestanding ERs, and other "sites of care." The chain was started in Nashville in 1968 by Thomas Frist, a cardiologist who felt "the hospital he worked in was poorly run and equipped." He and a few colleagues partnered with Jack Massey, the man who developed the Kentucky Fried Chicken chain. They built Parkview hospital in Nashville, then bought a second facility. When Frist talked to investors, he encountered skepticism. They didn't believe a doctor could be a good businessman. The secret of his success, he once said, was the idea of bigness itself.

By 2020, the company he founded had "180 hospitals and more than two thousand sites of care in twenty-one states and Great Britain." Its "market cap" exceeded $44 billion.

Another of the big four, Humana, was established after a few Louisville realtors figured they could make money operating nursing homes. They chose the company name from a list of five hundred submitted by a corporate identity consultant. After buying one, then a few, hospitals, they went public. By 1972, they were running forty-five hospitals. In 1978, they took over a second hospital chain and more than doubled in size. In 2018, the company had another growth spurt. This time they paid $4.1 billion for a 40 percent stake in seventy-six long-term acute-care and rehabilitation hospitals operated by Kindred Healthcare.

Bob Appel, founder of American Medical International, worked for a medical diagnostics lab that served hospitals. When his employer had financial problems, Appel purchased the lab and later started acquiring hospitals.

National Medical Enterprises, now Tenet, was formed by three lawyers in 1967. One was quoted as saying, "Doctors don't generally know how to run a business. A hospital is really just like a hotel. You just have to know the medical side."

Community Health Systems of Franklin, Tennessee, once owned twenty-five of the fifty hospitals that were the nation's most expensive. In 2014, they controlled two hundred hospitals and earned $18 billion in profits. Then they paid $7.5 billion for a for-profit chain "that had a slew of financial and legal problems," and they started losing money. When their debt was close to $15 billion, they sold thirty hospitals for cheap and they were teetering.

Currently, about 1,034 of the nation's hospitals belong to one of several for-profit corporations. They have stockholders, earn money, and pay taxes. In 2018, their bottom line was boosted (and government revenues fell) when the corporate tax rate dropped from 35 to 21 percent.

Strange as it may seem to some entrepreneurs, hospitals weren't initially established to create wealth. Our nation's first, Philadelphia's "Pennsylvania Hospital," was founded in 1751 "to care for the sick-poor and insane who were wandering the streets." The facility was the idea of a Quaker who was medically educated in Paris a century before physicians believed bacteria and viruses cause infectious diseases. He wanted to emulate Paris' Hotel-Dieu, the continent's second oldest. The Parisian facility was established by Saint Landry in AD 651 "to treat pilgrims and the poor." By the early 21st century, the facility had become central Paris' "top casualty and emergency hospital." It closed in 2013.

During the First World War, 116,000 Americans lost their lives and many were wounded. Four years later, the Veterans Administration was launched and they started building hospitals. By 2020, the VA was operating 170 medical centers and claimed to be the largest integrated health care system in the U.S.

The Indian Health Service is part of a federal agency and operates twenty-six hospitals that are directly and indirectly (through Medicare and Medicaid) federally funded. They help care for the descendants of people Europeans and others decimated and displaced.

After the American Civil War ended, 4 million former slaves were abruptly told they were free to leave, but they had no resources and nowhere to go, no basic shelter. Often cramped together in abandoned buildings, they couldn't maintain basic hygiene, and many got sick. Privately run institutions for the very poor existed, but there were no hospitals. The private shelters wouldn't accept the newly emancipated. Former slaves died in large numbers. In some places, "their bodies were littering the streets." At the time, there were 54,000 male and 300 female medical doctors in the U.S., and only a few were people of color.

During Reconstruction, Congress established the Freedmen's Bureau in the South. It "fed millions of former slaves and poor whites, built hospitals, and provided medical aid." But its hospitals didn't have enough beds, linens, or quarantine facilities and there was a smallpox outbreak.

In 1868, four years before Reconstruction ended, the government founded Howard University College of Medicine, an institution where hundreds of black students learned to become "competent and compassionate physicians who provided health care for the medically underserved."

In July of 1872, "responding to the continued hostility of white Southerners," Congress terminated the Freedmen's Bureau.

During the next ninety or so years, black physicians often couldn't get privileges to practice in hospitals dominated by whites. In many of our leading hospitals, people of color were cared for in the basement wards of white hospitals, and more than one hundred all-black hospitals were established. One of the facilities, Homer G. Phillips, was affiliated with my medical school. As a junior medical student, I spent three weeks as one of the two white-faced junior medical students who delivered babies.

The facility was named for a black lawyer from a small Missouri town who was raised in an orphanage, then by an aunt. He studied law at Howard University and practiced in St. Louis. In 1922, the city put an $87 million bond issue on the ballot. Phillips made sure the measure assigned one million of the dollars to the building fund for a black-run hospital. After the ballot issue passed, it wasn't easy to get the city to turn over the money, and Phillips spent more than a decade fighting "interests that sought to prevent the hospital's construction." The building's doors opened in 1937, and by 1961 "Homer G." had trained the "largest number of black doctors and nurses in the world. It was a leader in developing the practice of intravenous feeding and treatments for gunshot wounds, ulcers, and burns." It closed in 1979.

The hospital that provided a significant part of my medical training, St. Louis City Hospital, was taxpayer funded. It was racially integrated in 1955. When Medicare and Medicaid became the law of the land, it lost much of its clientele to more upscale facilities.

Cook County hospital in Chicago, the caregiver for the poor in our second largest city, is committed to providing "quality care with respect and dignity regardless of a person's ability to pay." It gets close to 40 percent of its operating revenue from state and city taxes. A third of the remaining 60 percent of the money comes from Medicaid. Medicare chips in 9 percent, and some of the people treated have private insurance.

New York City has eleven publicly funded hospitals plus clinics and nursing homes. With an annual budget exceeding $5 billion, they provide emergency services and hospitalize close to a quarter of a million people a year.

For people with complex or obscure problems, university hospitals offer a few extras. They are staffed by esteemed professors who teach, write papers and books, and are well read. Some of the professors survive on research money supplied by the NIH or pharmaceutical companies, and they participate in studies of new approaches or medications. The institutions earn some of their money because they handle medical problems that require a complex team approach. They are often the first to attempt interventions like liver, heart, and kidney transplants. They are also the training grounds for young doctors who are learning what to do and how to do it.

Close to half of the nation's 6,000-plus large hospitals are nonprofit, though many actually earn a lot of money. Some, like the Mayo and Cleveland Clinics, are nondenominational. A number are supported by Jewish and Protestant communities, and there are large hospital chains that were started and funded by one of the Catholic charities or religious orders. Many of these groups of hospitals have been merging with one another. A recent California antitrust suit alleges that when the hospitals in a geographic area are financially linked to one another, the cost of hospital care goes up. When the hospitals negotiate contracts with insurance companies, they don't price-compete with one another. During the last two decades, twenty-four of the large community hospitals of northern California became members of the non-profit conglomerate called Sutter. They started charging more for their services and demanding Health Insurance companies pay a higher a portion of the bill. By 2020, Sutter owned thirty-six surgery centers and employed twelve thousand physicians. The cost of private insurance rose and the California attorney general got involved. Calling the hospitals bullies and monopolies, he sued and Sutter settled. The company paid a fine of $575 million, altered the amount it charged out-of-network providers and got rid of some of their secrecy. In 2018 and 2019, Sutter had annual revenues of 12 to 13 billion dollars. In 2018, their medical operations netted over $100 million in profits, and in 2019 the conglomerate lost more than $500 million. The

TV show *60 Minutes* used the story as a partial explanation for the high cost of hospital care.

Medicare bases the amount of money it pays for each hospital stay on the diagnosis and the surgery or procedures that are performed. Insurance companies sign contracts with hospitals and more or less use the same approach. People without insurance or without the right insurance or without adequate insurance get hospital bills that are often wildly inflated. The charges, of course, are largely based on days spent in the hospital and procedures performed. But in addition, private hospitals like to bill for each "thing" that is used—each bag of fluid that is infused, each gown that is worn, and each bandage that is applied. The amount charged can vary a lot. A reporter for Kaiser Health News wrote of a Tampa hospital where, in 2008–2009, the bill for removing a gallbladder was as little as $48,000 and as much as $90,000.

Every hospital or chain of hospitals has a unique 65,000-item list of the amount they bill for each procedure, day in the hospital, or Band-Aid used. Until recently, the cost for each item was a deeply guarded secret. The Affordable Care Act tried to eliminate the secrecy. The law required hospitals to show their list to the public. Most hospitals didn't comply and the government had to issue a federal mandate that required hospitals to publish their catalog of charges on the Internet. It went into effect in January 2019.

The person who gets sick and is urgently admitted to a hospital doesn't always get to choose where they are treated. If the hospital doesn't have a contract with their insurer and the insurance doesn't cover out-of-network facilities, patients can receive a bill for their care. If the amount seems excessive, the people who were ill can negotiate, but they owe the money and there's no limit on the amount a hospital can charge.

A spokesman for "a nonpartisan forum called the national academy of state health care policy" wrote an article that defends high prices. He argues this is a free country and hospitals should charge what they want to charge. We shouldn't try to control the size of the bill as Medicare or insurance companies do. He doesn't say it's OK to overcharge the souls who are financially marginal and defenseless. And he doesn't say it's OK to turn every bandage, sheet, or gown a hospital uses into a commodity.

But he believes that in the land of the free and in the absence of a restrictive law, hospitals should be allowed to charge what they will.

Public and university hospitals derive funds from private and governmental insurance, from taxpayers, and from donors. The law says that if charitable organizations want to avoid taxes, they must be operated exclusively for "exempt purposes." None of the earnings can go to shareholders or individuals. Charitable organizations that want tax breaks can't try to influence legislation as a "substantial part of their activities." Some universities and charitable hospitals make a lot of money, and their profits are not taxed.

The Affordable Care Act did not challenge their IRS-exempt status, but it did add a few hoops. The institutions are now required to have a financial assistance policy that's written in "plain language" on their website. The "audience should be able to understand it the first time they read or hear it."

When a treated person doesn't pay their share of the costs, institutions are supposed to "make reasonable efforts to determine whether the individual is eligible for financial assistance" before they take legal action, sell a person's debt, or notify a credit agency.

In 2013, seven of the ten U.S. hospitals that made the most money were nonprofits. They earned over $1.5 billion, and, of course, didn't pay any taxes. They included a few big-name institutions: Gundersen of La Crosse, Wisconsin, cleared $300 million. Stanford's teaching hospital had a profit of $225 million, and the University of Pennsylvania's hospital in Philadelphia made $184.5 million. "The hospitals with the highest price markups earned the largest profits." These institutions could have returned some of the tax-derived revenues to the government. (Fat chance.) They also could have increased the number of non-paying patients they cared for. Since the Affordable Care Act went into effect, they're often caring for fewer people who can't pay.

A few years back, Senator Chuck Grassley considered "removing tax-exempt status from teaching hospitals and forcing them to do more for local low-income, urban communities." The hospitals fought him and won.

Many of the people who live far from the cities are served by smaller institutions that are the pride of their community but are financially struggling to exist.

In May 2020, the *New York Times* told how, as a result of the corona-virus bailout, the nonprofit Providence Health System received over $500 million in extra government funds. The company was sitting on nearly $12 billion in cash and in a good year was earning over $1 billion tax free from its investments in venture capital funds. Two other for-profit hospital groups, HCA and Tenet Healthcare, had billions of dollars in reserves and received $1.5 billion from the government. Ascension Health in St. Louis, a chain that had $15.5 billion in cash, was given $211 million. At the same time, the largest rural hospital system in eastern Kentucky got $3 million, enough money to pay their employees for two weeks.

"Before the current epidemic roughly four hundred hospitals in rural America were at risk of closing. An additional two thousand hospitals located outside towns and cities were in the black but on average only had enough cash to keep their doors open for thirty days." Sixty million Americans, 13 million of whom are children, live in rural areas. Facilities located in states that did not increase the number of people eligible for Medicaid were especially hard hit.

A third of the health care dollars, $1.1 trillion (2017), was spent on hospital care.

- Total Number of All U.S. Hospitals: 6,210
- Number of U.S. Community Hospitals: 5,262
- Not-for-Profit Community Hospitals: 2,968
- Investor-Owned (For-Profit) Community Hospitals: 1,322
- State and Local Government Community Hospitals: 972
- Federal Government Hospitals: 208
- Non-federal Psychiatric Hospitals: 620
- Other Hospitals: 120

In 1981, a deranged twenty-six-year-old man shot the president of the United States. A bullet hit a rib, and Ronald Reagan was in pain. At the time, neither he nor his aides sensed how close he was to death. As the presidential limousine sped to George Washington University Hospital, Reagan coughed up blood. When he tried to walk from the car to the emergency room, his legs collapsed. His aides kept him from falling and dragged him to a stretcher. A senior surgeon arrived on the scene and realized the man had lost a lot of blood and was still hemorrhaging. An operating room was empty and, as he was wheeled in, O-negative blood was pumped into Reagan's veins. The urgent surgery and transfusions were so successful that few realized Reagan had used up one of his nine lives.

The first U.S. level-one trauma center was established in Chicago in 1970. By 2003, most states had at least one facility where surgeons, neurosurgeons, orthopedists, appropriate anesthetists, and operating rooms were available twenty-four hours a day. Some facilities have helicopter landing pads. Others land planes in nearby airports and then use an ambulance to bring the severely injured to a hospital. Fifty million Americans are unable to reach one of these centers in less than an hour.

ERs in U.S. hospitals clock 136 million visits a year. Forty million of them are for minor injuries, auto accidents, and gunshot wounds, and 16 million of the people seen are admitted to the hospital.

People with myocardial infarctions who arrive by ambulance are immediately sent to the catheterization/stent department. Women in labor are wheel-chaired to the hospital delivery areas. People with a new neurologic problem—a leg or arm that won't move, inability to speak, loss of vision—are briefly assessed and routed to the CAT scan room. An X-ray without contrast only takes a few minutes, and doctors need to be sure the acute brain damage was not caused by a cerebral hemorrhage before they sometimes assume it was caused by the blockage of a vessel supplying blood to part of the brain, or start performing tests to try to discover the cause of the paralysis.

A few years back (according to a colleague), the physician-in-chief (PIC) of Northern California Kaiser was visiting one of our ERs and was disturbed by waiting rooms full of people. Some had minor injuries that needed a quick fix. Others were quite ill, but had to wait a long time before they were triaged. The PIC decided something needed to be done and he had clout.

He met with the chiefs of our hospitals and an agreement was struck. After a person entered the waiting room and registered, he or she had to be in an exam room within fifteen minutes. The sick person would then be evaluated by a doctor, nurse, and tech. Lacerations had to be sewn shut and minor fractures casted within an hour. Sepsis, strokes, and bleeding were treated promptly. When a person had a problem of unclear significance, tests and consultations were ordered. Efficient, competent evaluation might not always be possible, but that was the goal. A few years later, Dr. Pearl retired and waiting room wait time worsened.

The San Francisco General Hospital is a hybrid. It's a trauma center and a teaching hospital. The doctors who provide the care come from the University of California Medical Center. The facility also operates the city's "Community Health Network." Eighty percent of the hospital's $600-plus million operating budget is paid for by Medicaid, Medicare, and private insurance, and the facility gets the other 20 percent from the city's general fund.

A few decades back, "the general" developed a unique approach to people who were frequently treated in the emergency room and who slept outside in unsheltered locations. Many suffered and died prematurely from diseases caused by alcohol, smoking, exposure, and poor sanitation. Violence, mental disorders, and suicides were common. San Francisco General "identified individuals who had visited their emergency room five times in twelve months." With the help of attorneys and with a staff of case managers, a primary care physician, nurse practitioner(s), and a psychiatrist, they got many of the people permanent shelter, a primary care doctor, and "benefits." It was quite a legal feat for people who are disabled by mental illness. In time, 70 to 75 percent of the homeless were housed, and emergency department use decreased 50 to 75 percent. That was a few years back. Since then, the homeless problem in Northern California has gotten worse, and the funding has changed.

In 1963, I was the medical intern on call and was dozing when the nurse called and told me about a new admit who had come to the hospital with chest pain. An EKG showed he was having an acute heart attack, and he was told he could either get complete rest in the hospital or at home. He chose the hospital and was probably already asleep, but someone had to wake him and do the intake physical and the paperwork. At the time, there were no intensive care units.

I knocked on his closed door and got mentally prepared to meet a person who would be annoyed because I was bothering him. When there was no response, I opened the door and saw a middle-aged gentleman who was pale, sweating, and in shock. I called the nurse and we started an IV and infused the drugs that were available. He didn't respond and he was dead within the hour.

Nowadays, his occluded artery would have been ballooned open. A tube, a stent, would be splinting the part of the blood vessel that had been occluded, and he would be in an intensive care unit.

In the 1970s, hospitals established intensive care units for people with severe illnesses and those who needed a lot of attention. By the end of the '70s, each of the unit's nurses was responsible for no more than two patients. They used machines to monitor blood pressure, pulse, and blood oxygen levels. Respirators filled lungs. When appropriate, food was dripped into stomachs through naso-gastric conduits. Multiple IV tubes infused nutrients and a variety of chemicals and blood products.

Sometimes one of the occupants who was near death and had a problem that couldn't be fixed was kept alive on a breathing machine while the family prayed. Staying in the intensive care unit to postpone death for a few days is very expensive, but for some it became an expectation.

The glue that enabled a hospital to function as an organism, according to M.D. "poet" Lewis Thomas, was the nurse. He felt they were the people who best knew, talked with, and comforted the ill. They were the living human who was nearby night and day, and kept people clean, nurtured, well hydrated, and hopeful. He published his observations in 1974, and they still apply to most hospital nurses. But their chief job now is more mechanical and technical. They must make sure the IV doesn't run out of fluid, that blood pressure, pulse, temperature, and the level of oxygen in the blood is monitored, noted, and recorded, and that each ordered medication is given on time. Their eyes spend hours staring at a computer, and they are supposed to call the house doctor if something seems amiss. Too often, documentation seems to be their most important task. Every intervention and observation must be noted. If it's not in the record, "it didn't happen."

Nurses were traditionally underpaid. Their numbers increased during the Second World War, and by the end of the conflict, RNs earned an average of $2,100 a year, somewhat less than most male workers. When their wages leveled off, their numbers decreased and those who remained had to work longer and harder. By 1966, the average RN earned $5,200 a year. At the city hospital where I was a resident, one nurse passed medications and tried to handle the needs of an entire ward of sick people. The women (and at the time they were almost all women) were energetic and often exhausted. The wages at the City Hospital were so low that none of the nursing school graduates wanted to work at the hospital that trained them. When our few nurses went on strike in 1965, the papers thought their action was outrageous.

In 1981, the nation's air traffic controllers struck, and President Reagan declared they were federal employees and by striking they were breaking the law. He gave them 48 hours to return to work or be fired, and his anti-union stance challenged the nation's belief that unions have a right to strike. The following decade, major strikes plummeted from an average of 300 each year to fewer than 30. Despite an average annual inflation rate of 2.9 percent and a huge surge in productivity, the federal minimal wage, which was $3.35 in 1981, barely doubled in the three post-Reagan decades.

In the San Francisco Bay Area, two types of union actions are quickly resolved. The interruption of mass transit trains by striking workers leads to jammed freeways and lengthy commutes. The strikes seldom last for more than a few days. And a few decades back, when the police struck, the chief went on TV and warned that criminals, murderers, and rapists would soon have a field day. That walkout didn't last long.

At Kaiser, where I worked, nurses knew how to take care of the sick or needy, but they soon learned that dealing with management was something else. Every five to ten years, their contract came up for renewal and they asked for a raise. When management hung tough, the nurses felt they had to strike.

That was fine with Kaiser. The program was self-insured and the money kept coming in. Our patients didn't like to cross picket lines and they didn't come in unless they were quite ill.

Before the strike started, traveling nurses were hired, elective surgery was canceled, and the very ill were sent to other hospitals. I remember driving to work and seeing throngs of young and middle-aged nurses carrying picket signs and standing by the entrance of the parking structure. They were friends and colleagues. Many hated the thought of abandoning their patients, and some believed strikes were unethical.

After three weeks of pacing under an umbrella or in the heat of the day, nurses started returning to work. Most had loans, rent, bills, and lived from one paycheck to the next. When a majority of nurses had returned to the job, management offered a small raise and signed a long-term contract.

At the time there were two nursing organizations. One, the American Nursing Association (ANA), prohibited striking as a labor tactic, but strikes were not a problem for the California Nursing Association (CNA). In the 1990s the CNA recruited and hired a teamsters organizer named RoseAnn DeMoro. She was a good negotiator and they needed help. DeMoro accepted what she thought was a temporary job, but as one of her colleagues quipped, there's nothing more permanent than a temporary job. The CNA "disaffiliated from the ANA" and most nurses went with the California organization. Under DeMoro's leadership the nurses union started fighting Kaiser's anti-restructuring in 1997. Over an eighteen-month period, 7500 RNs at forty-seven Kaiser Hospitals and clinics engaged in eight short-term strikes. Each of the strikes was scheduled to last one to two days and the employer tried to fight back by locking the nurses out for five days. After the first day the nurses would show up at the hospital and say they were prepared to go back to work, but they were routinely turned away.

My memory of those days is quite limited, but the two day strike scenario went something like this: The hospital was given the required ten-day notice and they prepared. Critical patients were moved out, elective surgery was cancelled, and traveling nurses were hired. A few days prior to the designated date, the strike was sometimes held and sometimes cancelled and rescheduled for a date sixty days hence. A few weeks or months passed. As before, very ill patients were moved out, nurses brought in, and elective surgeries canceled. This time the nurses struck for two days and no nurse lost much money. As the strike was ending, the union announced it would soon strike again. Eventually management realized they would have to collectively bargain.

By April 1998, DeMoro's group had fought off the major Kaiser take-aways and had reached an agreement with the organization. The union felt they needed a militant approach because health care was rapidly becoming corporate and the change was affecting patient care, nurses' wages, and working conditions. The short-term strikes had a major economic impact and each strike was a major media event. They gave the nurses an opportunity to engage with the public, become patient advocates and spotlight issues at a particular hospital and/or corporation. The CNA gained a reputation as a fighting union and that helped them organize thousands of non-union nurses in California and nationally. In 1997 the California Nurses Association had 18,000 members. Twenty years later they were part of a national union with 150,000 members.

Nationwide, one in five nurses belongs to a collective bargaining unit. On average, their wages are 20 percent higher than the pay of the nurses who are not in unions.

Inside the large multispecialty hospital where I worked, there is a big birthing wing and large rooms accommodated by doulas and invited family. Nurse midwives handled the uncomplicated vaginal deliveries. An obstetrician always hung around and was available for consultation and urgent C-sections. An anesthesiologist or nurse anesthetist was always on duty and performed epidurals, infusions of lidocaine into the space between the vertebrae and the membrane that surrounds the spinal canal. The drug blunts much of the pelvic pain of childbirth.

In the hospital's basement, two CAT scanners and two MRI machines created detailed images far into the night. The pictures of the sliced body sections were immediately visible on every computer screen in the facility. A radiologist sat in a dark room, carefully checked the images, and looked for anything that didn't belong. He or she dictated a report that was transcribed by a trained worker in the facility, or increasingly by someone in the Philippines, India, or some other country.

A multi-bed outpatient recovery area was staffed by nurses. They prepared and observed patients before and after interventional radiologists performed biopsies or invasive procedures. The unit also served as the pre- and post-procedure area for people having an emergency heart catheterization.

A cardiologist was available twenty-four hours a day, to catheterize and stent the coronary arteries of people who were probably having an acute myocardial infarction.

Gastroenterologists, who sometimes worked at bedside in the intensive care unit, had a dedicated inpatient procedure room full of expensive equipment. When needed, we would insert a scope and treat people whose upper GI bleeding was ongoing or who had a bolus of food lodged in their esophagus.

There were medical, surgical, and pediatric intensive care units and a coronary care unit. Hospital rooms had piped in oxygen, suction, a place where a partner or parent could spend the night, and an annoying TV on the wall.

Pharmacists manned a special inpatient pharmacy twenty-four hours a day. They supplied a large range of medications, including specialized drugs and infusions.

Pathologists processed, stained, examined, and interpreted the significance of tissue that was removed from bodies. Samples were often sent "out" for molecular diagnostics (DNA/RNA analysis).

Phlebotomists drew blood and brought it to the lab where it was extensively tested. The labs also cultured and assessed stool, urine, blood, spinal fluid, and sputum.

Surgery accounts for a large portion of the hospital admissions and costs, and I dealt with some of the ins and outs in the chapter on surgery.

The 21st-century hospital is full of very costly equipment, commodities that are sold, serviced, refurbished, repaired, and periodically updated by well-schooled salespeople and technicians.

And there are clerks, social workers, discharge planners, janitors, patient transporters, painters, security people, guards, physical, occupational and speech therapists, dieticians, telephone operators, IT personnel, volunteers, educators, supervisors, administrators, billers, people who provide food for the patients and staff, and engineers who regulate and repair the electrical, cooling and heating, and other building systems.

In 2019, the Kaiser Family Foundation estimated that the basic U.S. hospital bed—excluding drugs, surgery, the fee charged by physicians, and much more—costs the providing institution $2,600 a day. Debt.org says the institution's cost is close to $4,000.

In 2009, a surgeon named Atul Gawande told the world about his visit to McAllen, Texas, a town that claimed to be the square dance capital of the world and whose health care was the second most costly in the country. It was near the Mexican border in a county where salaries were low. The townspeople didn't live longer than they did in nearby towns where the health care cost half as much. There was a lot of poverty, obesity, and heavy drinking.

The town had a number of "bread and butter doctors in private practice" who worked "from seven in the morning to seven at night and sometimes later." It also had a few large well-equipped hospitals and "virtually all the technology that you'd find at Harvard and Stanford and the Mayo Clinic." Gawande didn't directly answer the question he asked, but his article implied that McAllen's health care was costly because the doctors at the big hospitals performed a large number of expensive exams that were at the upper end of the procedures local doctors felt were appropriate.

Elsewhere on the planet, there are remote government-funded community hospitals, in low- and very low-income countries—like Malawi and Bangladesh. As described by the Harvard M.D. who spent time in Nepal, they are commonly located in far-off corners of their nation and are the only options available to the poorest of the poor. With 50 to 100 beds, they serve 100,000 to a million people, and their doctors can usually perform a few orthopedic and general surgical procedures and C-sections. They have some X-ray capability, but commonly lack certain basics. In Nepal, 15 percent didn't have piped-in water, 20 percent lacked electricity, 55 percent didn't have gloves, and 30 percent were unable to provide oxygen to those who need it.

CHAPTER TWENTY-FOUR

Generic Drugs

"Climate is what we expect. Weather is what we get"
—MARK TWAIN

In 1984, the Boston Celtics were NBA champions, Ronald Reagan was reelected president, and Ryan White, a thirteen-year-old boy with hemophilia, was expelled from school because he had AIDS.

It was also the year that Henry Waxman, a California Democrat, and Orin Hatch, a senator from Utah, transformed the world's approach to prescription drugs. Their new law allowed companies to produce and sell generics, copycat versions of approved medications, without having to again prove that the drugs did what they were supposed to do.

The companies that originally marketed the drugs had already performed the appropriate expensive double blind studies. Their studies had shown the medication performed the alleged task and that it was relatively safe. After the law was passed, repeat studies were no longer needed. The FDA was now expected to give a company permission to market a generic drug. The agency was "specifically prohibited from doing more than asking for bioavailability studies."

A federal agency with more than 22,000 employees, the FDA does much more than give marketing approval to drugs. It monitors a medication's side effects in humans and ensures "the safety, efficacy, and security of human and veterinary drugs, biological products, and medical devices." And, of course, it's responsible for the safety of our food supply.

The FDA became an independent agency that was responsible for the safety of the drugs we take in 1930s after a prescription drug killed more than one hundred people. The medication that caused the deaths was sulfanilamide, the world's first antibiotic. It was developed by the German company Bayer, and its patent was decades old when the company learned the drug "was effective treatment for a number of bacterial infections." At the time, sulfa was in the public domain. No one had to compensate the German company when they sold it. In the 1930s, a company in Tennessee created a sulfa elixir. They dissolved the medication in diethylene glycol, a compound normally used as antifreeze. They flavored the concoction with raspberry extract, saccharin, and caramel. It was a commercial success, but, as the company later learned, it was toxic and caused kidney failure and the death of over 107 men, women, and children.

There was a public outcry and Congress passed, and President Franklin Roosevelt signed, the Federal Food, Drug, and Cosmetic (FD&C) Act of 1938. For the first time, manufacturers were required to prove a drug was safe before it could be prescribed.

At the time, no one, not even the original creator of a medication, had to prove that a drug did what salespeople and hucksters claimed it did. Then thalidomide "happened" and tough rules were created.

Thalidomide was hailed by some as the tranquilizer of the future, a medication that gave people a night's sleep without a hangover. It was popular in other parts of the world, and many expected it to be a hit in this country. But the FDA kept delaying its release. Frances Kelsey, the officer in charge, had reviewed the application and she was dissatisfied.

Competent and a bit overeducated for her FDA position, Frances had, as she put it, "entered college in the depth of the depression and graduated when there were absolutely no jobs." Deducing she could either choose "to do graduate studies or join the breadline," she studied and acquired a master's degree. There still were no jobs, so she earned a Ph.D. In 1943, she was a biochemist at the University of Chicago and met and married a fellow biochemist. At the time, "two members of the same family could not be employed in the same department," so, she wrote, "Needing all the help I could get to obtain a job I entered medical school."

She was hired to be a medical reviewer for the FDA at a time when the agency was required to pass on a drug within sixty days or it would automatically be approved. Kelsey thought the thalidomide application was "incomplete." She learned that when the sedative was taken for a period of time, it sometimes caused peripheral neuritis, a painful tingling of the arms and feet. The effect had been recognized in Europe, and was the main reason the medication had lost its over-the-counter status in Germany. Within a year of thalidomide's introduction into Europe, a very rare deformity in newborn babies began to appear in Germany. It was called phocomelia. In the place of arms and legs, babies were born with something like fins. From 12 cases in 1959, the number grew to 83 in 1960 and 302 in 1961. Near the end of 1961, a Hamburg pediatrician made a statistical connection between this ominous health problem and mothers who had taken thalidomide while pregnant. The manufacturer was sufficiently concerned, and withdrew the drug from the market just as Israel was about to approve it. According to the FDA, ten thousand people in twenty countries were victims of the simple sedative.

Thalidomide was clearly effective, but it wasn't safe, and the nation's near miss made those who ran the country aware of a need to change the rules.

At the time, Senator Estes Kefauver, the "endlessly polite Southern senator in horn-rimmed glasses," was investigating "the escalating expense of lifesaving prescription drugs." He openly berated pharmaceutical executives for profiteering, and doctors were portrayed as dupes of the companies that produced our medications.

Known in the 1950s as the senator who investigated organized crime, the man from Tennessee campaigned for office wearing the kind of coonskin cap Davy Crockett wore when he died defending the Alamo. When Adlai Stevenson ran for U.S. president, Kefauver ran for vice president.

The thalidomide near miss gave him a platform, and he introduced amendments that gave the FDA the power to say it was no longer enough to prove a drug was safe in mice and rats. It now had to be shown effective. When his amendments became law, the FDA created hoops that were tough and the generic drug market dried up.

In 1984, the Hatch-Waxman Act made it easier for copycat drugs to come to market and that opened the floodgates for generic medications.

It also contained goodies for the pharmaceutical company that did the initial research on a new medication.

They were given a period of exclusivity—meaning they were protected from competition for the five years that followed FDA approval. Bruce Downey, the chairman of the Generic Pharmaceutical Association, once testified that the five years of exclusivity "was the greatest boon to pharmaceutical innovation in history because it forced brand manufacturers to replenish their products in the face of generic competition." The law extended the life of the drug's patents. Most were granted early in the research process and lasted twenty years. The legislation moved their starting point closer to the day when the medication was FDA approved.

It also dealt with the additional patents that companies file at various stages of the drug's development. These documents dealt with nonessential ingredients, like the color of a pill or the starches used as filler. The secondary patents were sometimes used as a pretext for a lawsuit. They allowed lawyers to allege a drug was still patent protected when its period of exclusivity expired. The legal assertion was often capricious, but the new law said that once the suit is filed, the FDA had to automatically delay the approval of the generic drug for thirty months "to permit litigation." For drugs that were very profitable, the extra thirty months added up to a lot of money. When a new drug earned a lot of money and its five years of exclusivity came to an end—when it was time to share the market with a generic competitor—some companies went low.

When lawyers filed the suits alleging patent protection, generic drug makers knew they would eventually win. They also knew that lawyers can drag things out. Court battles are lengthy and expensive. So, representatives from both sides often got together to make a deal. The company that owned the revenue-generating medicine paid the generic company a few million dollars a month. The generic drug maker agreed to wait for a year or more before it produced and marketed its product.

The Justice Department decided the practice was illegal and challenged the companies in court. In 2010, there were thirty-three pay-for-delay agreements, and in 2013, the Supreme Court ruled the practice was subject to antitrust laws. Two years later, the generic drug maker Teva settled a pay-for-delay antitrust lawsuit and paid $1.2 billion.

By 2012, 84 percent of the pharmaceuticals used by Americans were moving along the assembly lines of generic manufacturers. They were part APIs—active pharmaceutical ingredients—and part "excipients"—inactive substances.

In 2010, the U.S.'s top sources of pharmaceutical imports by "value"—not quantity—were Ireland and Germany. "In 2019, companies in India supplied 18 percent of the world's generic medications." China didn't sell many "finished drugs," but they produced 40 percent of the global chemical components used for our medications.

There are a number of generic manufacturers whose annual revenues exceed a billion dollars. The largest company, Teva, is based in Israel and has factories in many countries. In 2018, Teva had $19 billion in sales. Number two that year was Sandoz, the generics division of Novartis. They sold $9.9 billion worth of drugs, had 25,000 worldwide employees, and operated thirty manufacturing sites.

Some of the medications the companies made had "thin profit margins that sometimes led to shortfalls, manufacturing delays, or decisions to discontinue a drug altogether." When several companies manufactured a medication, competition could lead to a "race to the bottom"—a destructive drop in prices and shortages.

Pharmaceuticals are supposed to make money. When generic medications became less profitable, manufacturers often kept producing them in "older and less efficient production facilities." To update their facilities, some companies insisted on "predictability and incentives in the form of guaranteed-volume contracts to mitigate the risks of making investments." When contamination caused plants to close, manufacturers didn't always rebuild.

In December 2017, Teva faced falling profit margins and was under pressure in the U.S. from "major chains, wholesalers and benefit managers who had gotten together and demanded discounts." To survive and be profitable, they planned to discontinue some drugs and close or sell "a significant number of manufacturing plants in the United States, Europe, Israel and other markets."

In 2011, there were 257 new drug shortages, and in the 21st century, more than one hundred medications were hard to get "at any given time." Sixty-three percent of them were generic sterile injectable drugs.

When Obama was president, the FDA had four "officers" working on the shortage problem. The agency couldn't require manufacturers to do anything, but in 2015, acting on Obama's executive order, the FDA asked companies to notify the agency before they stopped making any or enough of a needed drug. The government wanted to be able to "determine if other manufacturers were willing and able to increase production."

The agency was told to bend a few rules: "Exercise temporary enforcement discretion for new sources of medically necessary drugs" and help affected manufacturers identify the reason there wasn't enough of the needed medication. They developed risk mitigation measures, like using sterile filters to allow the release of individual batches of a product that didn't initially meet the established standard. In 2017, "due to the ongoing lack of injectable drugs used in critical care," the FDA extended the expiration date of a number of Hospira injectable medications. In August 2018, the agency tested a stockpile of expired drugs that were owned by the military. Fifteen years after their stated expiration date, 90 percent of the one hundred prescription and over-the-counter medications "were perfectly good."

There's a gray market. Some facilities pay premiums and stockpile drugs and fluids. That's illegal, you say. Tell that to the person with a serious infection who needs the right antibiotic now.

"With the proper drugs," 90 percent of the three thousand children a year who are afflicted by T cell acute lymphoblastic leukemia can be cured. "Between 2009 and 2019, nine of the eleven medications doctors used to treat the disease were intermittently hard to get." In 2019, two companies were producing Vincristine, and one of them, Teva, stopped. One pediatric oncologist called it "a truly nightmare situation. Vincristine is our water. It's our bread and butter." Another doc added: "You either have to skip a dose or give a lower dose, or beg, borrow or plead." We will still be able to cure childhood leukemia, but it's harder "with one hand tied behind your back."

Bruce Chapner, an investigator at the National Cancer Institute, wrote about the limited availability of "workhorse cancer drugs" and the astounding short supply of many generic drugs: antibiotics, blood pressure meds, anesthetic agents, and electrolyte solutions.

Fifty-six percent of hospitals surveyed by the FDA reported that they had changed patient care or delayed therapy in light of scarcities. Thirty-six percent said they had rescheduled nonurgent or emergent procedures. "In Europe, the past five years have seen global shortages of at least ten essential oncology drugs. Two of five European doctors surveyed said the problems occurred on a weekly basis and typically last a few months."

As global demand increased, some shortages have become worldwide. Brazil had a three-year shortage of very long–acting benzathine penicillin G (BPG) at a time when the country was having an outbreak of syphilis. Benzathine penicillin G is a long-acting antibiotic that is known to cross the placenta and prevent mother-to-child transmission of syphilis. The disease is linked to severe malformation in babies. Worldwide, "just four companies produce the active ingredient for penicillin. The medicine offers little profit, and the companies keep production levels low."

According to the FDA, approximately twenty million IV saline bags are used per month in the United States. IV fluids are used to hydrate and as a vehicle for infusing medication. The U.S. gets a lot of its supplies from Baxter, a U.S. company with three manufacturing plants in Puerto Rico. (Baxter also has "twelve sites in the continental U.S. and eleven in Latin America and Canada.") Bags for the U.S. market are also manufactured by B. Braun Medical—the American branch of a German company that has operations in sixty-four countries, and Hospira (a Pfizer-owned company).

The supply of sterile solutions in the U.S. had been borderline for years when Hurricane Maria blew through Puerto Rico and knocked out the island's power. When I visited San Juan the following January, the warm Caribbean sun and the gentle island breezes masked the destruction. The functioning hotels were filled with construction workers, and there were no tourists. If someone in an apartment building had a generator, it was connected to electric wires from other apartments. Baxter asserted their three Puerto Rico sites "were minimally damaged." They said they had resumed "limited production activities within a week using diesel generators." The U.S. needed bags, and the FDA acted. That March they "checked out" Fresenius Kabi, Norway's saline-producing facility. It passed inspection and the agency "temporarily allowed Fresenius Kabi to distribute normal saline in the U.S." In April 2017, the FDA let Baxter temporarily import normal saline produced in Spain.

The world will soon face a shortage of the very special antibiotics that treat drug-resistant bacteria. According to the CDC, bacteria that aren't destroyed by current antibiotics infect at least 2.8 million Americans a year and contribute to 35,000 deaths. In the last decade, at least two start-ups received financial help from the government and Bill Gates, and they created medications that destroy resistant organisms. The drugs were approved by the FDA and are available, but the funding that helped them get on their feet ended. Doctors intentionally use the medications sparingly. We want to prevent or at least delay the development of resistant bacteria, but the antibiotics have become commodities. They have to earn their keep, and have value. At the time of this writing, the involved companies aren't selling enough and they are considering bankruptcy.

CHAPTER TWENTY-FIVE

The Price of Everyday Drugs

I n the U.S., doctors often don't know (and many don't want to know) how much people pay for their medications. Some people fork over the full list price. Others have health insurance that includes a drug plan and are only responsible for a co-pay.

In the world where everyone negotiates, these programs are managed by PBMs, prescription drug management companies. Some of these companies are owned by large health insurance companies. Three independent corporations control 75 percent of the market. "The negotiators receive a service fee from their clients, wrest discounts from drug manufacturers, and when appropriate share their financial gains with health plans." In the process, the managers determine which drugs insurers cover, and determine each medication's "tier" position.

To get a sense of the amount drugs cost people, I checked the online formulary of the University of Maryland Health Advantage. It has five drug levels, and five escalating co-pays. The following is the thirty-day charge for each filled subscription.

Tier one covers preferred generic drugs and the copay is $4.

Tier two is for generic drugs that didn't make the preferred list. Copay $15.

Tier three: preferred brand-name medications. Copay $47.

Tier 4: brand-named products that didn't make the preferred list. Copay $100.

Tier 5: 33 percent of the manufacturer's list price.

The top-tier medications are "specialty drugs" that typically cost $50,000 to $100,000 a year. Medications in the most expensive group usually include: combinations of anti-retroviral (HIV) drugs; agents that "modify" multiple sclerosis; a number of the very expensive, cancer-fighting medications. And orphan drugs—medications that are usually not in high demand for people over sixty-five. They only account for one percent of all prescriptions written, but they are responsible for 30 percent of the money Medicare D and Medicaid pay for prescription drugs.

Medicaid: Thanks to legislation sponsored by Senator David Pryor, the drugs Medicaid buys don't cost the government more than the drugs purchased by hard-bargaining insurance companies. Pryor is the son and grandson of sheriffs and was "arguably the most popular Arkansas politician of the modern era." In 1975, when he was the state's governor, his "frazzled wife ran away from the state's mansion and left her three sons in the care of her husband, setting Little Rock on its thoroughly Southern ear." She went to school for two years, and produced a feature-length film, a "kind of a witchcraft western." Then she returned to her family "as a complete person."

As the chairman of the Senate Special Committee on Aging, Pryor believed "the high cost of prescription drugs was one of the biggest problems burdening seniors." He held hearings and "attacked drug industry leaders." Then he decided to help Medicaid—the government program that provides coverage for 64 million of the nation's poor and disabled and covers the cost of nursing homes for many. Medicaid is funded by the state and federal government, with the feds on average paying 50 percent of the cost in wealthier states and up to 75 percent in states with lower per capita incomes.

In the late 1980s, many states were in financial trouble. They tried to limit the use of prescribed drugs by people on Medicaid by creating "restrictive formularies, co-pays, and monthly maximums." At the time, states were paying full sticker price for prescribed medicines while insurance companies and the VA were often given discounts of 30 to 40 percent. Two states tried to bargain with pharma and were attacked by the industry. Pharma argued that if states withheld "brand-name drugs without generic equivalents from a Medicaid enrollee," they would be endorsing "second-class medical treatment for the poor." In the late 1980s, President George H. W. Bush and his White House staff decided to "shrink the budget deficit" by about $50.5 billion. The legislation they produced was "massive"—533 pages long—"the five-year Omnibus Budget Reconciliation Act (OBRA 1990)." Its size and scope allowed Pryor and colleagues to add their "Medicaid Prescription Drug Rebate Program" to the bill. It granted Medicaid "most-favored customer" status, and required drug manufacturers to sell their medications to Medicaid at the "best price" available to any other purchaser. If a company wanted their products to be covered by each state's Medicaid prescription program, they had to accept the federal pricing provisions.

Medicare D: Thirteen years later, the controversial law that got the federal government into the drug business was passed in the early morning hours of June 27, 2003. The law gave drug "benefits" to people who were insured by Medicare, and it was called Medicare D. It didn't restrict a manufacturer's right to charge what it thought it could get away with. Quite the opposite, it prohibited the secretary of HHS (Health and Human Services) from bargaining for lower drug prices.

When asked why he thought House leaders had scheduled the vote long after most Americans had gone to bed, Representative Dan Burton (R-IN) said, "A lot of shenanigans were going on that night (that) they didn't want on national television." According to Walter Jones, a disgusted North Carolina Republican who voted "no," it was the "ugliest night" he had witnessed in more than two decades as a member of Congress. "Pharmaceutical lobbyists wrote the bill."

Touted as a means of providing cheap or free drugs for people on Medicare, the bill did not include any new taxes. The entitlement was not funded, though part of the cost was paid by enrollees. Seniors paid $265 to receive the benefit, and then kicked in $25-plus a month. When the annual cost of a drug exceeded $2,400 a year, the government stopped paying for a while. Enrollees had to pitch in for the next $4,000 worth of medications, and the government started paying again if the annual drug cost exceeded $6,400. The price was a "catastrophe" The $4,000 was called the donut hole, and it was eliminated by Obamacare.

After the bill was passed, the Government Accountability Office claimed that American prescription drug prices rose 6.6 percent a year between 2006 and 2010. By contrast, the price of generic pharmaceuticals increased by 2.6 percent annually, and overall medical costs rose 3.8 percent a year.

"Representative Billy Tauzin (R-LA.), the 'Cagey Cajun'—he came from a French Speaking Louisiana family—coauthored the bill." "Then he negotiated a $2-million-per-year job as a lobbyist for the drug industry's trade organization." "Thomas Scully, a Bush Medicare official who misstated the program's cost, became a health industry lobbyist."

Medicare D didn't trigger a huge protest over the high prices of some of our most expensive commodities because an earlier law protected most of the people on Medicare from extreme drug costs and forced the government to pay most of the bill. In recent years, the law has also protected pharmaceutical manufacturers from the wrath of the nation's elderly.

According to the 1988 law, if the yearly outlay for a drug exceeds $6,350, the situation is deemed a "catastrophe." If a drug costs more than $6,400 a year the Medicare recipient is then only responsible for a small copayment. The legislation essentially authorized the government to pay the other 80 percent of the annual cost of expensive drugs. But that wasn't its intention. It started as a bill that limited "total out-of-pocket charges for people on Medicare." Otis Bowen was its champion. He was a country doctor from a town of five thousand in northern Indiana. He was also his state's governor before he was appointed secretary of Health and Human Services. Known as "Doc" Bowen, he "kept a prescription pad handy and recommended remedies to cure the common cold and sore throat for both colleagues and members of the press." When his bill to help people on Medicare was introduced, President Reagan wasn't keen on the idea. But he was in the midst of a huge scandal. His administration was caught red handed giving weapons to Iran and money to the Central American counterrevolutionary fighters known as "Contras." Senator Lloyd Bentsen made sure the legislation didn't cover most medicines. It was just for catastrophes. At the time, there weren't many mega-expensive drugs. But the law was passed over thirty years ago, years before the enactment of Medicare D. Over the decades, the cost of specialty drugs has gotten quite high. Between 2010 and 2015, the amount Medicare Part D spent on these medications nearly quadrupled. They were costing the government $8.7 billion a year in 2010 and $32.8 billion a year in 2015. That year, they accounted for 31 percent of the program's net spending.

Senator Bentsen brokered the compromise that led to the government paying most of the cost of very expensive drugs. He was a fearsome poker player. You never knew what cards Bentsen held. During the Second World War he was a B-24 bomber pilot and was shot down twice. Looking and dressing like Hollywood's version of a successful politician, the senator was tall and thin, had a deep voice, and wore elegant clothes. When he went to a party and his wife wouldn't leave, he famously would "playfully toss her over his shoulder like a sack of potatoes and carry her out." In 1988, when Michael Dukakis was the Democratic presidential candidate, Bentsen was his running mate and they lost.

Outside the U.S., nations negotiate, buy the drugs, and create formularies. Germany, for example, makes sure the medication is safe and effective. The country then allows the manufacturer to set a price and sell it. During its first year, the new drug is compared to existing therapies and its relative "value" is determined. At the end of the first year the company and government authorities bargain. They usually arrive at an acceptable price. If they disagree, they arbitrate. The panel of five that hears both sides includes a representative of the government and a person who speaks for the insurer. Their decision is binding. If the manufacturer strongly disagrees, they can choose to stop selling the drug in the country. In seven years, Germany has assessed "more than 300 drugs and fewer than 30 were withdrawn."

CHAPTER TWENTY-SIX

High-Priced Drugs

The last two decades have witnessed an increase in drug prices most years. The amount charged for newly released specialty drugs has increased dramatically. Medicines that combat cancer, multiple sclerosis, and arthritis currently cost $60,000 to over $100,000 a year, and ads on TV promote the benefits of a number of expensive monoclonal antibodies.

Once the FDA approves a medication, the pharmaceutical company that owns it has a five-to-twelve-year head start on the competition, a period of exclusivity that is granted by the government. Manufacturers set the price of their medication and their lawyers patent just about everything about the drug they can think of. Humira had 126 patents. Politicians like to complain about the high cost of medications, but few seem to be doing anything about them.

On February 26, 2019, the heads of seven pharmaceutical companies appeared before a congressional committee that was trying to learn why medicines cost more in America than they do in other advanced countries. During the hearing, senators seemed to be asking industry leaders for something they could bring to their voters. Early on, Debbie Stabenow of Michigan concluded that companies keep raising prices and charge more because they can. She did not, however, note that prices are also high because company leaders are expected to take advantage of the loopholes and incentives that were written into the laws.

The CEO of Merck, Kenneth Frazier, told the senators that he felt his company had a duty to be responsible in pricing practices and to contribute to solutions that address patient affordability. The son of a janitor, and grandson of a South Carolina sharecropper, Frazier was a Harvard graduate and the first black man to head a Fortune 500 corporation.

The head of the French company Sanofi, Olivier Brandicourt, "spent two years as a young doctor in the Republic of the Congo. He studied malaria during his eight years at the institute of infectious and tropical diseases in Paris." He said his company had pledged to keep price increases at or below the U.S. national health expenditure projected growth rate. He also mentioned the gap between net and list prices. Drugs cost more, in part, because of the cut that middlemen and -women take.

Albert Bourla, the new CEO of Pfizer, was a Greek veterinarian who was a former director of the company's animal health group. He told the senators that he had made it clear to investors: Pricing will not be a growth driver for the company now or in years to come.

AstraZeneca's leader, Pascal Soriot, called Australia his home. He kept a road bike in the Alps, another in the U.S., and a third in Cambridge, England. He pushed the cost problem back onto the senators and said, "[If the U.S. doesn't like the way we price drugs,] the government has to step up and change the rules."

The CEOs of the two Swiss giants, Roche and Novartis, both of whom have a significant presence in the U.S., were notably absent.

In less developed countries, local patent laws once allowed people to purchase a number of generic medications relatively cheaply. In 1995, the World Trade Organization (WTO) was formed, and nations who wanted to trade with the big guys were forced to join. The WTO championed the concept of "intellectual property" and forced the "less developed countries" to change their drug patent laws. At the time, many countries allowed pharmaceutical manufacturers to patent the way drugs were made but not the drugs themselves. After 1995, the actual medications had to be patented and the patents had to last for a minimum of twenty years.

That obviously threatened the health and welfare of many of the world's poor. Many of the people who were about to see a worsening in their access to needed medications struggled to have enough money to buy food. They would never have enough money to buy high-priced patented drugs. Forcing their country to abide by strict WTO patents would never remotely affect the earnings of wealthy Western pharmaceutical companies. But the rules were the rules. The leaders of the WTO were expected to defend the principle of "intellectual property."

At times the WTO punted. They delayed the onset of the new rules so nations could "reorganize and restructure." When the waiver period was ending, Bangladesh, a country that was "nearly self-sufficient in pharmaceuticals," appealed. They said they were "struggling to provide their population with prevention treatment and care. The new patent laws placed many critical treatments outside their reach." The drugs that cured hepatitis C in six weeks, for example, were priced at $1,000 a pill in the first world. They were available for the poor in Bangladesh at $10 a pill. The WTO granted Bangladesh an additional "waiver" but ruled that by 2033 the country either has to comply with the new patent laws or leave the organization.

CHAPTER TWENTY-SEVEN

Expensive Pharmaceuticals

The Swiss

It's hard to say when the price of newer drugs started spiraling upward, but a Swiss medicine called Gleevec—a "small molecule" that attacks malignancies in a unique way—probably played a major role.

The medication didn't appear on the scene until I was practicing for more than a dozen years. In the late 1950s, some people with advanced cancer were treated with nitrogen mustard derivatives. During the subsequent decades, a number of drugs that attacked rapidly growing cells were developed. In the 1960s, doctors started using combinations of these medications to cure some lymphomas, leukemias, and a few uncommon malignancies—like some metastatic testicular cancers and choriocarcinoma.

The chemicals caused major side effects. But when they were taken by people with widespread cancers, tumors shrank and some lives were extended. The drugs were later used to destroy "probable" metastases, cancers that had seeded parts of the body but were not visible or detectable. Doctors determined the odds that they existed and were hiding in the body by studying the surgically removed primary.

Gleevec was the product of decades of research at the Ciba-Geigy lab in Basel, Switzerland. It was discovered by a research team chasing a dream, a theory, a hypothesis. Alex Matter, a Swiss M.D., advocated looking for a small molecule that would get inside cancer cells and stop them from growing.

"Inspired by the likes of Louis Pasteur and Marie Curie," Matter was twelve years old when he began dreaming that he would one day be involved in the "discovery of important new medicines." In 1983, he became a Ciba researcher in Basel, the centuries-old city that straddles the Rhine just before the river bends and enters Germany. He presumably wondered why the offspring of a normal cell kept reproducing and didn't die when it was supposed to. Was one of its numerous tyrosine kinase enzymes at fault?

Each part of the body is made up of cells. Within each of these small units, traffic is directed down various metabolic pathways by enzymes called kinases. At the appropriate time, they cause cells to "grow, shrink, and die." Matter wondered if one of the proteins that function as an "on" or "off" switch got stuck in the pro-growth position. Could that be the reason abnormal cells don't die and keep growing? What if we could block a corrupting kinase without harming the other ninety or so known tyrosine kinases? Could we cure that cancer? That was the dream.

Kinases have inlets on their outer surfaces. When these are filled by a small molecule that "fits," the cell dies. Locating the bad kinase and plugging it with the appropriate small molecule is a little like finding a needle in a haystack. But that's what the Swiss Geigy team lead by Alex Matter and Nick Lydon set out to do. In 1986, someone discovered a small molecule that selectively inactivated one and only one of the ninety or so known "tyrosine kinases" that were known to be present in each cell. Matter made various small alterations to the protein, created new molecules, and tested them one by one. A few seemed promising and the researchers gradually accumulated dozens of kinase blockers. The project took years and must have been quite costly. Geigy funded the studies "reluctantly." Matter was told to keep investigating other approaches to cancer. The kinase program was supposed to be "very, very small—hidden in plain sight."

In the 1980s, Lydon went to Boston in search of a cancer that might be susceptible to one of his kinase inhibitors. He met Bryan Drucker, a lanky, soft-spoken physician from St. Paul, Minnesota, who had spent nine years at the Dana Farber Institute. Drucker was studying chronic myelocytic leukemia and was interested in Matter's drug. Matter was unable to provide the chemical because lawyers for the Swiss company and the Dana Farber Institute "could not find agreeable terms."

During the next few years, the drug remained in Switzerland. Drucker moved to Oregon, contacted Lydon, and requested the drug. When he visited the Oregon university lawyers and asked for permission to test a Swiss drug, his request seemed casual. No big deal. The university lawyers "humored" Drucker and signed.

After he obtained the kinase inhibitors, Drucker tested them. The blocker caused chronic myelocytic leukemia (CML) cells in a petri dish to die. Then Drucker planted tumor cells in a mouse and gave the mouse the drug. This time the tumor got smaller.

Most cases of CML are caused by a genetic accident. The tips of two chromosomes have "broken off," switched location, and fused. The resulting "hybrid" gene, called the Philadelphia chromosome, causes the abnormal cells to keep reproducing. The defect that causes the disease had been elucidated and explained a decade earlier by a hematologist named Janet Rowley.

In the absence of a bone marrow transplant, chronic myelocytic leukemia was usually lethal. If a person had an HLA identical sibling, and if they underwent a stem cell transplant, they subsequently had a 60 to 80 percent chance of surviving and being disease free five years out. Without someone else's bone marrow, half of those affected were dead in three years, and fewer than one in five lasted ten years.

But a dream was being realized. A small molecule could selectively inhibit an enzyme and control or cure cancer. Turning the protein into a drug a human could use required proving its safety in animals, then people. Several hundred million dollars needed to be spent before the company could market the medication. It would only help a few thousand people.

Novartis, the company created by the merger of Ciba-Geigy and Sandoz, decided to give the chemical a shot and see what it did to the cancer in question. The first clinical trial of imatinib mesylate (Gleevec) took place in 1998, and Gleevec was amazingly effective. In 2001, the FDA approved the new medication and granted Novartis a five-year monopoly.

When it first came out, Novartis worried because most cancers eventually become resistant to therapy, but they were pleasantly surprised. Gleevec and a slightly altered later iteration "changed the natural course of the malignancy." It didn't cure chronic myelocytic leukemia, but it turned it into a chronic disease, a problem people were able to live with.

Each year, an additional group of people developed CML and started taking a pill a day for the rest of their lives. A 2015 study of people who had taken Gleevec for ten years found that "82 percent of them were alive and progression-free. By 2018, "an estimated 8,430 people in the United States" were living with the diagnosis.

When it was first marketed in 2001, the CEO of Novartis called Gleevec's introductory price, $26,000 a year, "high but fair." Over the next few years, the company started increasing the price in parallel with "the purchasing power of money." After 2005, yearly boosts started exceeding inflation by 5 percent, and by 2007, Gleevec was costing consumers "$3,757 a month ($45,000 a year)." Its list price exceeded $60,000 in 2010, and it passed the $100,000 a year mark in 2013.

When the first generic form of the drug entered the U.S. market, it wasn't much cheaper, but it could be purchased in Canada at a third the U.S. price. In 2015, the U.K. and France were paying out about "$30,000 per person per year and Russia was spending $8,370 for each individual."

In 1970, India was still allowing drug companies to patent their manufacturing processes but *not* the active chemical or the drug. A number of relatively cheap generic medications were developed and marketed. When the country joined the World Trade Organization, India started allowing companies to patent the drugs themselves, and in 1998, Novartis filed a patent for a new form of Gleevec. The Indian court decided the company didn't prove that the modified Gleevec worked better than the form that was being sold in India. It accused the company of "evergreening," extending the life of a medication by altering it just enough so they could get a new patent, and it rejected the Novartis claim.

To keep doctors from viewing Gleevec as just another commodity, to avoid having to compete on price alone, Novartis researchers looked for a molecule that was as effective as or better than Gleevec. They modified the protein and developed nilotinib (Tasigna), a chemical that did a better job at targeting the kinase in question. It rescued some people whose disease no longer responded to Gleevec, and it more rapidly and effectively reversed the biochemical markers of chronic myelocytic leukemia in new patients. But it was not better than Gleevec at halting disease progression—and it wasn't worse. In 2007, the FDA released Tasigna, and a few years later, a Canadian study showed that 5 to 6 percent of people taking the new drug had a heart attack, stroke, or some other "atherosclerosis-related ailment." When it learned of the drug's dangers, the FDA didn't ask Novartis to remove the medication from the market. The company had to put a black box warning on the drug insert.

Seven years after nilotinib was approved, a year's worth of the medicine was costing over $100,000 and the company was expecting revenues in excess of $2.5 billion.

After Gleevec started making a lot of money, pharmaceutical researchers started looking for other kinase inhibitors. In 2013, the FDA approved ibrutinib, a kinase inhibitor that improves the survival of people with chronic lymphocytic leukemia. Two years later, AbbVie paid $21 billion for the company that developed the drug. To the delight of their stockholders, AbbVie charged over a hundred thousand dollars a year for the medication. A few years later, a Dutch company opened a research facility in Northern California. In 2017, they got FDA approval to sell a similar kinase inhibitor called acalabrutinib. It was priced at $168,000 a year.

As the money from the sale of Gleevec accumulated, Novartis grew in size and almost became as profitable as its Basel compatriot Roche. The Swiss companies are, in terms of revenue, two of the world's largest. In 2019, Roche, at number two, was taking in more than $63 billion a year while Novartis had a gross income of over $44 billion and was the world's fourth largest.

In 2006, Novartis planted its feet solidly on American soil when it paid $5.4 billion and bought the 51 percent of Chiron they didn't already own.

The business they purchased was twenty-five years old and was created by Bill Rutter, a visionary biochemistry professor who started the company and used genetic engineering to help create a vaccine for hepatitis B.

Chiron's importance grew in large part as a result of the discoveries of a research team led by Michael Houghton. His group sought the identity of the mysterious virus that was present in some units of transfused blood and caused hepatitis. Houghton was a Brit. He came to the U.S. in 1982, and with two colleagues, worked in the labs of Chiron for seven years. They studied the blood of infected chimpanzees, "followed false leads, dealt with repeated frustrations," and eventually found a small piece of nucleic acid. Using it, they were able to clone the virus. Once they identified hepatitis C, the company developed a blood test for its presence and we learned that 70 million people were chronically infected and 20 percent of them had cirrhosis.

Years later, Houghton, the "relentless and honorable" man who had led the project, moved to Alberta, Canada, and became a professor. A number of decades later he was offered the Gardner Award in recognition of the epic nature of his discovery. The prize was one level below the Nobel Prize. Houghton insisted that he would only accept the award if he could share it with his co-discoverers. They had been excluded a few years earlier when he accepted the Lasker, the highest U.S.-based award in the life sciences. He didn't want to relive the inner conflict caused by his acceptance of the prior award. The Gardner committee refused to honor his coworkers and Houghton "agonized." Deciding he didn't want to do "that" again, Houghton refused the award and it was given to someone else. In 2020, he was one of three hepatitis C researchers who received the Nobel Prize in medicine. His collaborators were again excluded.

In 1984, Chiron researchers worked on the first sequencing of the HIV genome and developed blood tests and vaccines. In their early years, the company participated in joint ventures with Ciba-Geigy and at some point Chiron's leaders sold 49 percent of the company's ownership to the Swiss. Ciba-Geigy later merged with Sandoz and became Novartis. Representatives from the two companies met every six months in Basel or Emeryville, California, and they discussed the progress of various ventures. On one occasion, the meeting was in California. The locals wore jeans and dressed casually while the Swiss arrived in business suits. At the meeting, a California executive arrived in a flower tank top and shorts. The Californians smiled as they watched the Swiss stiffen. In 2006, **Novartis** bought the other 51 percent of Chiron and the company started wearing the mantle of a U.S. corporation

After Novartis was in charge, they marketed the first U.S. biosimilar, Zarxio. It's a medication that stimulates the bone marrow to produce more neutrophilic white cells. In 2018, Novartis paid $8.7 billion for the therapeutic gene that is used to treat the lethal genetic disease called spinal muscular atrophy. As mentioned elsewhere, they plan to charge $2.1 million for a course of therapy. More recently, Novartis developed a company that plants cancer fighting genes into T lymphocytes. The process is called CAR-T.

The CEO of the other Swiss pharmaceutical giant, Roche, was an Austrian-born economist. Married with three children, he skied, hiked, and made movies in his spare time. Initially thought of as shy, he headed the company when it bought California-based Genentech for $46.8 billion. The cost of the merger virtually cemented Roche's need to charge high prices and to sell a lot of their drugs.

After the "merger," the companies also had to deal with a "clash of cultures between a freewheeling Californian biotech company and a buttoned-up Swiss multinational." There was plenty that could go wrong. The California innovator "was full of smart people who were very upset and worried about the idea of another company coming in and making the decisions."

Genentech had started as a company that used genetic engineering to produce hormones like insulin and TPA, a protein that dissolves clots. At the time of the purchase, Genentech owned three antibodies that were used to fight cancer: bevacizumab, herceptin, and rituximab.

One of the big three, an antibody called Avastin, prevents the growth of the new blood vessels that a cancer needs if it wants to get bigger. In 2010, the expensive commodity generated $7.4 billion in revenue for its new Swiss owner.

The concept of a gene that stimulates the growth of the blood vessels that nourish malignancies was around for years before the drug was developed. The idea was initially developed and promoted by Harvard surgeon

Judah Folkman. As a boy, Judah had accompanied his rabbi father when he visited people in the hospital. "His father would pray through oxygen tents and Judah would sit in a chair and be very quiet. About age seven to eight he noticed doctors could open the tents and do things, and he knew he wanted to do things. He told his father he wanted to become a doctor not a rabbi. He feared his father would be upset, but his dad replied you can be a rabbi-like doctor, and Folkman felt better."

Folkman served in the navy for two years, went to med school, and became a surgeon. In the 1950s, "he developed the first implantable pacemaker that targeted the region between the upper and lower heart chambers." He also "pioneered the first polymers that could be planted under the skin and slowly released drugs. When he was thirty-four, Folkman was "the youngest ever Harvard professor of surgery." He had a research lab and he studied the blood supply of tumors. By 1971, he had learned that "the growth of solid neoplasms is always accompanied by vigorous new capillaries that come from the host." Time-lapse movies of an animal experiment demonstrated vessels advancing toward and penetrating a tumor implant and establishing blood flow. When the new vessels didn't develop, most solid tumors stopped growing when they were two to three mm in size. Tumors didn't die but the growths became inactive. Folkman's lab isolated a factor that stimulated rapid formation of new capillaries in animals. His people unsuccessfully tried to develop an antibody to the factor. When he lectured to other doctors, Folkman promoted the idea that cancer enlargement could be slowed if we could block the stimulus for blood vessel growth. Over the years he noticed that when he rose to speak at medical meetings some doctors in the audience filed out. They knew what he was going to say. Some thought his idea was far-fetched, and others were tired of his pitch. Folkman would later quip that science goes where you imagine it. He believed there's a fine line between persistence and obstinacy.

In 1989, a Genentech investigator isolated and cloned three isoforms of "vascular endothelial growth factor" (VEGF), a gene that caused blood vessels to grow. Then a researcher named Napoleone Ferrara developed the first of many antibodies to VEGF.

Ferrara was born in Catania. It's a coastal Sicilian town that is 30 kilometers from the highest volcano in Europe. Ferrara's interest in science was ignited by his grandfather, a high school science teacher who had a five thousand–book library. In those days, the size of a person's library was an indication of how intelligent he was. The Sicilian went to medical school. He decided to become a researcher after he heard the fascinating lectures of a charismatic professor of pharmacology named Umberto Scapagnini. Ferrara joined Genentech in 1988. He and his group spent years characterizing the protein and developing the humanized antibody that stopped the growth of blood vessels. It became the drug Avastin. Ferrara was lecturing in Sienna when he learned that a pivotal study proved his antibody helped treat colon cancer. He recalled celebrating by drinking a whole bottle of Chianti.

Avastin remains pricey and is not always covered by insurers. Using it can create an additional burden for people who are living on a tight budget and have widespread disease. In 2008, Roy Vagelos, the former chief executive of Merck, commented on the rising costs of medications. He spoke about a drug that "costs $50,000 a year and adds four months of life. The unnamed drug he focused his criticism on was thought to be Avastin. He called it a "shocking disparity between value and price," and he said the high prices charged for Avastin were "not sustainable." His comments were quoted in the *New York Times*.

Keeping the price of Avastin high has been a struggle. In 2015, the British National Health Service and some insurance companies were disturbed by the thought of spending tens of thousands of dollars for the extra months of life. There was a general effort by many European countries to lower the price of their expensive cancer-fighting drugs. "A bid to push down drug prices by the Swiss health ministry allegedly infuriated drug makers, and the company warned that such a move would hurt employment and would have a negative impact on their future contribution to the Swiss economy." In the years subsequent to its release, Avastin's annual revenue always topped $5 billion.

The second drug Roche acquired, Herceptin, was also an antibody. Most cancer-causing genes "are sequestered deep in the cell." The gene in question, neu, is connected to the cell membrane and "a large fragment of the gene hangs outside." It was discovered in the 1970s when a researcher (working with Robert Weinberg at MIT) injected the "DNA from rat neurological tumors, into normal mouse cells. The injected cells turned cancerous." After the gene was discovered, it was "more or less forgotten," and largely ignored before one of Genentech's scientists, Axel Ulrich, made an antibody that targeted it. When Ulrich's antibody attached to neu, it created a complex. The assemblage was "noticed" by macrophages, white cells that "engulf and rid the body of cellular debris." The cell would sense the antigen-antibody combination, know it didn't belong, and obliterate it.

Once created, the antibody to neu might have intrigued some people, but it was not really useful. Ulrich talked about it in 1986 when he gave a seminar at UCLA. One of the attendees, Dr. Dennis Salmon, was interested.

A university hematologist, Salmon grew up in a coal mining town and, as a boy, had been impressed by the doctors who came to the house to tend to his father. His dad survived two mine cave-ins and lost a leg in an auto accident. The doctors making house calls "made people feel better." Salmon "saw the respect his parents gave them" and always thought being a doctor would be "pretty cool." In high school, he "developed a keen interest in biology." Then, in college, he spent summers working in a steel mill. The job was tolerable for a few months, but the experience showed him what life as a factory worker could be like. It "cemented his resolve. This wasn't what I wanted to do with my life." After med school, Salmon had job offers, but he accepted a position at UCLA because "It wasn't ossified. If you had some resources and a good idea, you could pursue it."

According to Mukherjee, Salmon attended Ulrich's lecture on neu and thought he and Ulrich should collaborate. The Genentech researcher gave UCLA a DNA probe that identified neu, and Salmon checked his array of cancer samples to see if any of them were, perhaps, driven by the gene. Until that time, the gene had only been found in mouse brain tumors. There didn't seem to be much chance that it would turn up in a human tumor.

But it did. The oncogene, now called Her-2/neu, was found in some breast cancers. It turned out to be an important reason for their rapid growth. Some breast cancers made and used it in large quantities. Scientists implanted Her-2-containing cancers in a mouse and watched them increase wildly. Trastuzumab, the antibody that inactivated Her-2, caused the cancer cells to die.

The scientific findings were intriguing, but it took a while before Genentech was fully committed to the idea of making a cancer drug. It would be a first for them. A drug that interfered with cancer was still a reach.

Salmon kept working on the project. He couldn't use the standard mouse monoclonal antibody because it could trigger an immune response. They found a Genentech scientist who knew how to create a mouse that produced monoclonal antibodies that a body would think came from a human. In the summer of 1990, they successfully created Herceptin. Women with breast cancer became experimental subjects. In 1992, fifteen women were studied, and in 1996, nine hundred were given the drug. It kept making a difference. In 1998, the drug application was submitted to the FDA and it was quickly approved. Its initial monthly price was $3,208, and in 2013 rose to $4,573.

The overall cost of creating a new drug was significant. Testing, development, and getting FDA approval cost a lot. I suspect hundreds of millions of dollars were spent in the process. But the reward, $6 billion-plus a year, dwarfs the expense and adds to the culture of high drug prices and the commoditization of cancer medications.

When Roche announced their revenues in 2016, rituximab, the third Genentech cancer-fighting antibody, topped the list. With $7.3 billion in annual sales worldwide and $3.9 billion in the U.S., the drug was on fire.

Some people believe medications should be treated like any other commodity. The companies that produce them should charge as much as they can get away with. Investors should be richly rewarded. Others believe illness can strike anyone, rich or poor, at any time, and the cost of health care should be shared. To mollify these people, the drug industry tries to justify the prices they charge by invoking the cost of developing a new drug. They don't usually supply details.

Rituximab provides a window into how much it really costs to create an innovative medication when researchers have a strong sense of where they are going and how they are planning to get there.

Approved by the FDA in 2012, the injectable antibody has revolutionized the treatment of some lymphomas. It targets CD20, a unique protein that is found on the surface of only one kind of human cell, the B lymphocyte. The mice that make the antibody were genetically altered, and the antibodies they make are hybrids—chimeric—part mouse and part human.

After rituximab is infused, it circulates and "tends to stick to the side of B cells that are rich in CD20. Natural killer cells then destroy up to 80 percent of a body's B cells."

The drug was developed by a San Diego start-up called Idec. Its founders included several Stanford university researchers and Ivor Royston, a San Diego immunologist. Royston was the son of a Polish sheet metal worker who entered Great Britain via the beaches of Dunkirk. Royston fondly remembers the summer when he and his mother lived in the castle his father was re-roofing. The building had once been the home of Henry VIII and Anne Boleyn. In 1954, the family moved to America and Ivor was a good student. He went to medical school, and married a woman whose father was a successful businessman. His father-in-law liked to "challenge" young Ivor with business problems. If the son-in-law couldn't solve the problem, his father-in-law would tell the young doctor how stupid he was. Years later when he was running Idec, Royston "wasn't afraid" to get involved with business people. "If I could deal with my father-in-law, I could deal with anybody."

After medical school, Royston carried out research at the NIH and became board certified in oncology. He tried "to understand how the body recognizes cancer cells, and searched for a way to encourage the body to have an immune reaction to them." When he was a low-level research doc at Stanford, Royston was stirred when he learned how to make monoclonal antibodies. "You could produce antibodies by fusing lymphocytes with myeloma cells and create a cell that doesn't die and keeps making antibodies." A colleague went to England, contacted the physicians who discovered monoclonal antibodies, and brought back cells from the "the myeloma line, the immortalizing cell line." He gave a few of the precious "hybridoma cells" to Royston, and Ivor spent the next twenty-two years trying "to figure out how to make antibodies against cancer cells."

From the outset (1985), Idec researchers were looking for a monoclonal antibody that could be used to treat B-cell lymphomas. Their efforts consumed millions of dollars. There are about 240,000 cases of lymphoma in the U.S. each year, and we now know the antibody they developed is also useful in people with autoimmune and inflammatory diseases.

In 1991, the company needed additional funding to continue. They had an initial public stock offering and sold some of the business. The proceeds netted enough money to get through FDA phase one—toxicity and dose—and phase two—treating patients without a control group to see if the drug seemed to work—testing. The company had allegedly spent $80 million to this point. They did not have the money necessary to perform phase three, a double blind study comparing the drug with an inactive substance. The FDA wouldn't approve a drug until the study was performed.

In 1995, their CEO, a former Genentech guy, signed a collaboration agreement with his former employer. The giant chipped in $60 million and acquired "a majority of the sales and profits that Rituxan would generate if it earned FDA approval."

Rituximab was initially approved in 1997, and out of the gate, Genentech charged $3,475 for a month's worth of the infusion. In 2002, $1.47 billion of the drug was sold. Genentech got most of the money and Idec got $370 million. By 2013, the average thirty-day cost of infusions had gone up to $5,031.

Vis-à-vis the price having something to do with the cost of development, Idec spent $80 million and Genentech spent $60 million, for a total of $140 million. In 2017, the medication brought in over $7,000 million—$7 billion. The three antibodies Roche acquired when they bought Genentech accounted for more than half of the company's revenue. In 2017, the Swiss company owned three of the world's most expensive pharmaceutical commodities. They sold $7 billion worth of Avastin, $7.4 billion worth of Herceptin, and $9.2 billion worth of Rituxan.

The Best-Selling Drug in the World

In 1999, as part of a hostile takeover, Pfizer paid $90 billion and swallowed Warner Lambert, the corporation that owned/controlled Lipitor.

In a twelve-year span (1997–2009), Pfizer sold more than $80 billion of the pharmaceutical for $5 to $6 per twenty-mg pill.

To understand the drug's appeal, we have to remember that 800,000 deaths a year—one in three deaths in this country—are the result of a heart attack or stroke. Ninety-two million adults are living with heart disease or the aftereffects of a stroke. Vascular disease isn't preventable, but it's more likely to occur in people with a family history, high blood pressure, diabetes, and in those who smoke or have a high serum cholesterol. Since the 1950s, doctors believed that if they could lower the level of cholesterol in the blood, they could prevent some heart attacks. A low-fat diet helps, but major dietary restrictions sometimes fail.

Our bodies require cholesterol. It's a component of cell membranes, hormones, and much more. We mainly produce it in the liver. When we make too much, our blood level rises and we are at risk for heart disease. Over a few decades, researchers have unsuccessfully sought a molecule that would block the synthesis of cholesterol. The chemicals they came up with were toxic or not very effective.

In 1973, a researcher in Japan, Akira Endo, extracted a statin from a blue-green mold, and it caused a human's blood cholesterol level to drop. As he tells his story: "I was born into a rural farming family in northern Japan, in Akita. I lived there for seventeen years with my extended family, including my grandparents. My grandfather, who had an interest in medicine and science, was a great home teacher to me. Thanks to his influence, I became fascinated with mushrooms and other molds. At the age of ten, I dreamt of becoming a scientist." Akira was also inspired by a biography of Alexander Fleming, the man who discovered penicillin, the juice that protected a blue-green mold. It made him wonder what other amazing juices do these fungi fabricate? He became a researcher and spent two years in the U.S. He was surprised by "how much richer the American diet was relative to the diet of the Japanese." While living in the Bronx, he recalled seeing ambulances that took people who had suffered a heart attack to the hospital.

When he returned to Japan, Endo became an investigator for a Japanese pharmaceutical company. He looked for fungi that could produce antibiotics or affect the synthesis of cholesterol. His company turned him loose, and over a number of years, he studied the "juice"—the metabolites and toxins produced by over six thousand types of fungi. In the early 1970s, he found a product of a blue-green mold that blocked cholesterol synthesis. It was growing on a rice sample in a grain shop in Kyoto, and it became the world's first statin. Others modified Endo's discovery and developed increasingly better cholesterol lowering medications.

After confirming Endo's findings, Merck scientists used an aspergillus (and Endo's methods) to create the cholesterol-lowering drug lovastatin. A 1994 Scandinavian study showed that statins "led to a sharp drop in fatal heart attacks for patients with heart disease." The following year, Merck sold more than 4 billion dollars' worth of simvastatin and lovasatin, and a number of researchers tried to make a superior statin. Bruce Roth, a scientist who worked for Warner Lambert in Ann Arbor, synthesized Lipitor. He showed it was very potent—the strongest available. Roth had grown up outside Philadelphia, loved the night sky, and initially wanted to be an astronomer. One day when he was in high school, reality struck. He "concluded there were only like three jobs in the whole world for astronomers," and he decided to become a chemist. While learning the craft, he came to believe that "there really are true artists in organic chemistry." He got his Ph.D. and played intramural softball at Iowa State University. At age thirty, he moved to Ann Arbor, Michigan, and "used his skills" to create the statin molecule called atorvastatin.

After it was tested and FDA approved, Warner Lambert joined forces with Pfizer, and the drug was "aggressively" priced and promoted. Then (apparently) something went wrong with the business relationship. Warner sought to be acquired by someone other than Pfizer, and Pfizer wasn't having it. In 1999, as part of a hostile takeover, Pfizer paid $90 billion and swallowed Warner Lambert. The company went on to sell a lot of Lipitor. When Pfizer's monopoly ended, their profits dropped and the company attempted to avoid U.S. taxes by moving their headquarters to Ireland. In response, the Obama Treasury Department eliminated the tax loopholes that would have saved Pfizer money, and the company called the move off.

The Commodity You Could Barely Give Away

When Roy Vagelos was head of Merck, the company "vowed to only increase prices in line with the Consumer Price Index, plus or minus one percent and about half the industry followed suit." When some companies used loopholes in the drug laws to extend the patents of their successful drugs, Vagelos refused to join in.

Vagelos was the physician and academic lipid researcher who became the company's CEO in 1966. The son of immigrant Greeks, he worked in the family restaurant when he was young and he played his violin in his free time. Roy's grandfather had been a doctor in the old country and had died young. Roy's father felt that to succeed in life his son would need a good education. Vagelos studied hard and went to college on a scholarship. He always recalled how he was snubbed by the Johns Hopkins interviewer. The meeting had been going well but ended abruptly when the interviewer learned the young man's parents had not gone to college. Vagelos was clearly not Hopkins material.

As a university student, Vagelos developed a love for chemistry. His first year in medical school was a "very tough time because he had a terrible memory. Anatomy almost wiped him out." Fortunately, there was also biochemistry and he "survived" and became a talented researcher.

In 1966, he became the chairman of the biochemistry department at Washington University in St. Louis, and in 1975 he became the head of Merck. The company had started in Germany in the 1800s and opened its U.S. branch in 1891. In its early days, the company made medicinal morphine and codeine. It was the birthplace of one of the first medical books for the masses, *The Merck Manual.*

Under Vagelos' leadership, Merck developed lovastatin and simvastatin, the first drugs that limited the body's production of cholesterol. While Vagelos was in charge, a Merck research team led by William Campbell developed a drug that killed a number of the parasitic worms that attacked cattle, sheep, and horses The man who led the effort was a transplanted Irishman. As a schoolboy during the Second World War Campbell had walked to school carrying a gas mask in a little box slung around his shoulder. His Merck group launched their efforts after a "fermentation broth" eradicated worms from an intentionally infected mouse. The medication they developed, ivermectin, was marketed as a means of preventing heartworm in dogs. It didn't do much for the hookworm-like parasites that lived in the intestines of man, and its commercial value seemed limited. Further research on the chemical was suspended, and it was shelved. One day, Mohammed Aziz, a staff researcher, met with Vagelos and got permission to perform additional studies. Aziz had been in Africa and had seen people who had been infected with the filariae that caused river blindness. A hundred million Africans were at risk for the condition and the parasite had blinded 18 million of them. The invading worm existed in two forms: adults, which can be six to fifteen inches long and exist as lumps under an infected person's skin, and the filariae, a small organism that infiltrated the skin and caused intense itching. The black fly that lived in the river spread the parasites from one person to the next.

People who had the problem were constantly scratching themselves. When kids scraped their skin then touched their eyelids, the microfilaria got into their eyes. The subsequent inflammation led to scarring and blindness. In some villages, 25 percent of the inhabitants couldn't see. In an attempt to escape, many moved away from the river to less fertile ground and suffered from malnutrition.

Ivermectin was a Merck drug that destroyed the filariae that attacked horses, but it had been one of the large pharmaceutical company's financial failures. Aziz suggested it might have an effect on the creatures that blinded so many Africans, and Vagelos was on board. Merck produced a quantity of pills, and Aziz went to Senegal to study their effect. Pinch biopsies of the skin of infected people showed huge numbers of the filariae. Half of the people who were infected got a pill and the other half didn't. A month later, a second biopsy showed the filariae had been eradicated from the people who had been treated.

Based on the positive results, Merck spent years performing tests that proved ivermectin was safe and effective. Then they went to the African leaders and tried to sell it for a dollar a pill. The government had no money. The discussion went something like: "OK, $0.50 a pill, a dime." The governments really didn't have enough money. The World Health Organization was spraying rivers with insecticides (though the black flies were already becoming resistant to the spray). The WHO wasn't interested. Officials in the U.S. State Department and at the White House were excited, but "the government was broke." (Ronald Reagan was president.) The French were about to approve the drug. (There were cases in Paris that had originated in colonial Africa.) In the U.S., the FDA wasn't helpful.

Merck was in business to make money and to enrich its officers and stockholders. But the drug was ready. These were the 1980s, and Roy Vagelos was a doctor as well as a businessman. The leadership at Merck decided it wasn't a commodity. It had no economic value, but they wanted to provide the medication free of cost to anyone who would use it. They had spent millions to develop the medication. Providing it gratis would cost the company (and its shareholders) tens of millions of dollars, but Vagelos made the announcement and waited to see how the stockholders would react. He claims he received a lot of positive feedback but didn't get one negative letter. For years thereafter, the best of the best researchers in the country wanted to come to and work for Merck, and Vagelos stayed on as head of Merck for an additional six years.

When Vagelos reached the mandatory retirement age in 1994, Merck "was number one in sales, size, and marketing force." As a successor, Vagelos recommended a number of pharma-savvy colleagues, but the world was changing. The board chose a real businessman—a non-scientist, Harvard MBA, and former CEO of a medical device company named Ray Gilmartin.

Thirteen years after Vagelos retired, the Columbia Business School sponsored a great debate on the ethics of drug pricing and Vagelos was a major participant. He felt the "drug pricing problem" started in the early 20th century when biotech companies started marketing "targeted treatments" for cancer and other diseases. He recalled "almost choking" when he heard Genentech was charging $50,000 for Avastin, a cancer drug that lengthened a person's life for three to four months. At the same time, he believed in the marketplace and "accepted" the company's right to charge what it thought it could get away with. He concluded: "The industry has a lousy image and it should, until it reforms itself."

Twenty-five years after Merck started giving ivermectin to millions of the world's poorest, two of the drug's creators received a Nobel Prize. One of them, William Campbell, was an Irish researcher who worked for Merck. The other, Satoshi Omura, came from a farming area near Mt. Fuji. Omura's parents believed in conformity and tradition. They taught him, "deru kugi wa utareru ga, desugita kugi wa utarenai" (a nail that sticks out will be hammered down, but a nail that sticks out a lot will not). The grandmother who helped raise him, however, "repeatedly emphasized that for personal development and satisfaction, it was always best to work for the sake of others." After graduating, Omura taught evening classes at a Tokyo high school. Then he joined the Kitasato Institute and spent his life isolating and evaluating bioactive substances that microbes produced. Avermectin, the starting point for ivermectin, came from a Streptomyces in the soil at the edge of a Japanese golf course. Omura isolated it, and his institute gave the drug to Merck. The pharmaceutical company "handled the animal testing, development, production, marketing, and distribution."

One of the World's Most Valuable Commodities

With annual revenues of almost $20 billion, Humira is currently one of Earth's most valuable commodities. More importantly, it is a monoclonal antibody that blocks inflammatory cytokines, small proteins that play an important role in our body's defense against attacking viruses and bacteria.

Our skin and intestinal tract protect our bodies from a world full of bacteria, viruses, and other microscopic creatures. If an invader gets past the barrier, they will almost always be detected by one or many of the immune cells that float through our blood and lymph. These cells identify, imprison, and destroy pathogens. In the process, some people get sick. When or if a defensive system fails, some people die.

Our immune cells communicate with one another and coordinate their attack by releasing small proteins called cytokines. There are at least five families of these molecules. They play a variety of roles and like snow-flakes they look a little different from one another. In their exuberance to eliminate an invader, an overproduction of cytokines can worsen the pneumonia caused by COVID-19 and certain other organisms. There's a new kind of cancer treatment called CAR-T. When it works, it can suddenly destroy large numbers of malignant cells. That can produce a large outpouring of cytokines. The response is called the "cytokine release syndrome," and it can cause diarrhea, heart damage, and lung injury.

In more chronic situations, defenders can mistake good guys for bad guys, and the cytokines can attack joints (rheumatoid arthritis), the intestine (Crohn's), the kidneys (lupus), or the nervous system (multiple sclerosis).

Many of today's biologics are pricey monoclonal antibodies that block inflammatory cytokines. In 2017, biologics in general accounted for "2 percent of U.S. prescriptions and were responsible for 37 percent of net drug spending."

TNF—tumor necrosis factor—is the misleading name of a family of cytokines that is a major cause of the pain, swelling, and inflammation suffered by people who have any of a number of autoimmune diseases. The protein's name was chosen by researchers who were trying to understand how some malignancies were cured when they were intentionally infected with virulent bacteria.

The story of the TNF cytokines and the creation of antibodies that block their action started in the 1960s and '70s when a number of scientists developed mouse myeloma cells (malignant plasma cells) that could be grown in tissue culture and that, along with their progeny, survived indefinitely. They were said to be "immortal."

Then a researcher named César Milstein, who was working at Cambridge, learned how to turn two small myeloma cells into one larger cell. Milstein was Jewish and was born in Argentina. His father had exited Russia when he was 14 years old. His mother was born in Argentina and was the headmistress of a school. She encouraged her son to study hard and go to the University of Buenos Aires. At one point, she helped type her son's Ph.D. thesis.

After he graduated, Milstein married and spent three years as a postdoc fellow at a lab in Cambridge, England. In 1961, he returned to Argentina and was supposed to head a department at the university, but the politics of the country had changed. A military coup had taken control of the government, and it was conducting a campaign against political dissenters and targeting Jews. When he was an undergraduate, Milstein had taken part in demonstrations against the country's president, Juan Peron. He felt unsafe and returned to the lab at Cambridge and worked there. In 1974, he started working with Georges Köhler, a Swiss postgraduate researcher who was also interested in fusing myeloma cells.

That year, the two Cambridge researchers injected purified protein into a mouse. One of its lymphocytes recognized the invader and started cloning, making copies of itself. Each new lymphocyte predictably produced antibodies to the same foreign protein. After a few days, one of the researchers drew blood from the animal's spleen. As expected, the serum was full of identical lymphocytes, each of which produced the same antibody. So far, nothing exceptional had happened.

At this point, the researchers attempted to fuse some of the lymphocytes to mouse myeloma cells, malignant plasma cells that don't die. The combination was supposed to keep reproducing and keep making quantities of one and only one antibody.

As Köhler, a shy, gentle Swiss German immunologist, later explained, the fusion approach was new and unique. "If by blind chance the right lymphocyte, the one producing the antibody against the injected antigen, had fused with the myeloma cell and it was forming daughter cells that were locked into producing the same pure antibody . . . it was a long shot." Around Christmas 1974, Köhler added the antigen to the fused antibody-producing cells and went home. If the experiment worked, the antibodies would combine with the antigens. They would precipitate, and halos would form around the cells. Köhler returned hours later, and, fearing failure, brought his wife along to console him. They looked in the window, saw the halos, and were elated. "I kissed my wife. I was all happy."

The two researchers wrote a paper and told the world about the first "hybridoma"—part lymphocyte, part myeloma cell—the first "factory" that produced monoclonal antibodies. In the tradition of Howard Florey, the developer of penicillin, and Albert Sabin, the creator of the oral polio vaccine, the researchers did not patent their creation.

A few years later, a researcher named Hilary Koprowski, "a colorful, prominent Polish-born virologist," used the monoclonal antibody procedure in his research and decided to patent the method.

Once he owned the process, Koprowski and an entrepreneur named Michael Wall formed a company named Centacor, and they tried to figure out how to turn mouse monoclonal antibodies into gold.

In the summer of 1982, Michael re-met Jan Vilcek, a man who worked at a New York hospital and studied cytokines. A Czech researcher, Jan was a six-year-old Jewish kid in 1938 when the British prime minister "gave" his country to Hitler in return for a promise of peace. During the next few years, the Nazis rounded up and killed Jews. Vilcek and his parents survived "because they had a complicated attitude toward their Jewishness." At some point, they converted to Catholicism, changed their address, and Jan's father joined the underground. One way or another, they managed to avoid the death camps.

After the Second World War, Russia took control of Czechoslovakia, and the country became part of the Eastern Bloc. The Soviet Union and the U.S. feared one another and built nuclear missiles. They maintained armies, restricted travel between the Soviet Bloc and the West, and forbade immigration.

Vilcek married and became an accomplished researcher. Two years after the first known cytokine, interferon, was identified, he started studying it. The cytokine got its name because it was secreted by infected cells. Interferon then made nearby cells resistant to the virus.

When he was in his twenties, fed up with the Czech Communist government, Vilcek decided to "relocate." In 1964, he and his wife received permission to cross the iron curtain for a three-day vacation in Vienna. It was October and still warm. They traveled by auto and brought their heavy winter coats with them. When they reached the border, their car was searched. Jan worried that the coats would be a giveaway—that the inspectors would realize that he and his wife were trying to escape. He waited while the border guards "hesitated for the longest minutes of his life before letting them pass." Once across the line that divided the countries, Vilcek and his wife didn't go back. After the couple reached Germany, life was rough, but within a year, Vilcek was hired by NYU, New York University.

After spending a number of years studying interferon, Vilcek attended a workshop on a poorly understood immune regulator, a cytokine called tumor necrosis factor (TNF). In 1984, Genentech scientists determined and published the complete amino acid composition of the TNF cytokine. They purified the human protein and gave some of it to NYU. Vilcek and his colleagues accepted the gift and "felt like kids in a candy store—what should we try first?"

The cytokine turned out to play a role in a body's ability to fight viral infections. It had so many additional actions that one of Vilcek's graduate students quipped, "TNF should stand for 'too numerous functions.'"

After cytokines are secreted by a cell, some bind to receptors on the surface of other cells and regulate the immune response. They sometimes work alone, sometimes together, and they can even work against one another.

In the 1970s, scientists learned how to alter the genetic makeup of mice. Genes are sequences of DNA. When a gene is injected into a fertilized mouse egg, the inserted DNA can become part of the mouse genome. When it is then passed from one generation to the next, it can instruct the mice to make certain proteins.

In the '80s, Centacor (still struggling), on a whim, a hope, produced a monoclonal antibody that would block or inactivate TNF. It was made in a mouse that was genetically altered so its antibodies were "chimeric," part human and part mouse. The development took experts at Centacor six months. The antibody's patent was owned by Centacor and NYU (an independent private research university). The cytokine blocker didn't (as Centacor had hoped) help people with sepsis. But it blocked the action of TNF, prevented a cascade of pro-inflammatory cytokines, and hindered inflammation.

When London doctors (Feldman and Maini) injected the medication into the swollen, inflamed joint of a person with rheumatoid arthritis, it helped for three months. A repeat injection was also successful. In 1993, a physician from Holland successfully used the antibody to treat a desperately ill twelve-year-old girl with a severe case of Crohn's disease, a chronic inflammation of the small and sometimes large bowel. The antibody was called infliximab and was approved by the FDA in 1998. It is currently a commodity, one of the treatments that are used to treat ulcerative colitis, rheumatoid arthritis, ankylosying spondylitis, and various manifestations of psoriasis.

Part of the research was funded by the NIH (the taxpayer) and part by Centacor. There was a lot of luck and serendipity along the way. Both Centacor and NYU were rewarded. In 1999, Johnson & Johnson bought Centacor for $4.9 billion. Revenues from the drug (per J & J) rose annually, and topped $7 billion in 2016.

Though the cytokine blocker helped, it was still immunologically part mouse. Industry felt it had to create a rodent whose antibodies were indistinguishable from those made by man.

Humira—adalimumab—is an antibody that blocks TNF. It was created in mice that were genetically modified in embryo to make antibodies that a person's immune system thinks are human in origin. Some of the research on Humira was performed by researchers at the government-funded Cambridge Antibody Technology, U.K.

When the drug was approved by the FDA and was financially successful, scientists in many of the world's labs knew how to make monoclonal antibodies to TNF. They couldn't market them without performing placebo control studies that proved their drug was both safe and effective. That was costly, ethically questionable, and medically unnecessary.

Early in 2010, President Obama signed the Biologics Price Competition and Innovation Act of 2009. It allowed companies to market "biological products" that were demonstrated to be highly similar (i.e., biosimilar), and basically interchangeable with an FDA licensed biological product. The biosimilars did not need to undergo double blind control trials before they were approved by the FDA. But the original drug's period of exclusivity was twelve, not five years.

When Congress debated the proposed legislation, some argued that it takes a long time to develop a new biologic, and giving a product extra years will spur innovation. The dozen-year monopoly was opposed by generic manufacturers and consumer groups. They argued that twelve years meant biologics would likely be expensive and often unaffordable for too long.

Bruce Downey, the chairman of the Generic Pharmaceutical Association, argued the long period would negatively affect innovation. In much of Europe, biosimilars can be marketed as soon as they have been shown to perform as well as the original drug, but they are usually not thought of as interchangeable or a generic version of an existing medication. They are instead treated like a competitor or a therapeutic alternative. Outside of a small number of Eastern European countries, pharmacists can't substitute a biosimilar for a designated medication without permission from the prescribing physician. Currently in France and Italy, and perhaps elsewhere, "biosimilar adoption rates have been low."

Several cytokine families (including interferons) contain both pro- and anti- inflammatory molecules. Drugs that block cytokines are currently available to molecules that belong to one of two groups: "TNF—tumor necrosis factor" and "interleukins."

High levels of a number of cytokines have been found in the blood of people with infiltrates in both lungs and low levels of blood oxygen caused by SARS, MERS, and COVID-19.

By late 2020, thirty-six interleukins had been identified, and pharmaceutical researchers had developed, tested, and marketed humanized monoclonal antibodies that block some of them.

Using a clause in the law to pressure competitors, the owners of Humira extended its twelve-year U.S. monopoly for an additional five years. Their action might have been the real reason for the congressional inquiry that was held on February 26, 2019.

The session in question was officially an investigation of drug prices. The senator chairing the session was Ron Wyden of Oregon. He focused his questions on Richard Gonzalez, the CEO of the company that makes and markets Humira.

Gonzalez had worked for Abbott Laboratories for 30 years. He was the current head of AbbVie, the Abbott spin-off that owned Humira. In 2007, he developed throat cancer and retired. His cancer was successfully treated, and when he recovered, he decided he would "go back and do something." His company's main moneymaker, Humira, was FDA approved in 2002, and it was given a twelve-year monopoly. During that time, other manufacturers realized the drug was a gold mine and they developed generic biosimilars, drugs that worked as well but were chemically a little different.

When the twelve-year monopoly drew to a close, AbbVie lawyers allegedly used "a patent thicket of overlapping and add-on patents as well as the litigation process itself, as a mechanism to protect its blockbuster." The AbbVie legal team was prepared to initiate a court battle that could keep competitors off the market, but they started negotiating with the companies that had spent years and perhaps millions of dollars developing biosimilar medications. In the end, the competition was allowed to market the biosimilars in Europe and other parts of the world. They agreed to avoid the U.S. market until 2023. They also consented to pay royalties to AbbVie when they sold their drugs in the U.S. Watchdog groups thought the agreements were illegal and wanted the Federal Trade Commission to investigate.

At the 2019 Senate hearings, Gonzalez did his best to avoid discussing the biosimilar elephant in the room. He talked about his company's failed attempt to develop a drug that cured hepatitis C and stated a willingness to work with the committee. He pointed out that his company is charitable and explained that it has thirty thousand employees (and they need to eat). Senator Wyden, in turn, pointed out that Gonzalez's salary was $22.6 million and he had recently received a bonus of $4.3 million. Wyden asked if the bonus was tied to the financial performance of Humira.

(At this point, if I were Gonzales, I'd say I didn't write the law. As chief, I'm supposed to hire lawyers who will exploit the loopholes you guys allowed our lobbyists to put into the legislation. If you don't like what's happening, change the law.)

In September 2018, Congress passed John Sarbanes's Biosimilars Competition Act. It is supposed to "shine a light" on backroom, so-called "pay-for-delay" deals that were often made in secret. They have to be reported to the Federal Trade Commission who, with the Justice Department, is supposed to review agreements, look for antitrust and anti-competitive behavior, and "punish bad actors." In other words, if a company has a drug with annual revenues of $10 billion and they keep competitors off the market for five additional years, their company could bring in an additional $50 billion. Then the naughty owner has to notify the government, admit no wrong, and pay a fine of a few million dollars. Great law.

The Commodity a Texan Refused to Let Pharma Ignore

"Discovery consists of seeing what everybody has seen, and thinking what nobody has thought."
—ALBERT SZENT-GYORGYI

That's what Jim Allison had to deal with when he learned how to use the immune system to attack and sometimes cure metastatic cancer. Allison was from a small town in Texas. His mother died of lymphoma when he was still a boy, and his father was a country physician. He initially planned to follow in his father's footsteps, but he got interested in research when he was a high school student. In an interview, he said that he was reluctant to become a physician because doctors have to be right almost all the time. Researchers, on the other hand, develop hypotheses and test them. If they aren't wrong most of the time "they're not on the edge."

He wrote poetry, liked to read, loved country music, and later in his career played harmonica on the stage with Willie Nelson. At the University of Texas, he studied biochemistry and earned a Ph.D. After his post doc year, he got a job in a small lab that the University of Texas/M.D. Anderson Cancer Center was operating in Smithville, close to Austin.

While in Texas, he worked out the structure of the T cell antigen receptor and gained some notoriety. T cells are one of the white blood cells that float in our blood and are part of the immune system. They aren't very good at recognizing abnormal proteins, but they are efficient destroyers. When a dendritic or other watchdog cell spots a virus, it processes the "stranger." Then it "presents" it to the antigen receptor on the T lymphocyte" and the T lymphocyte takes over and deals with it. At one time early in Allison's education, he recalls a professor who doubted there was such a thing as a T cell.

After his "receptor" accomplishment, Allison took a year sabbatical and became a professor at the University of California, Berkeley. He considered himself an immunologist. His lab tried to work out the relationship between the T cell and cancer. In animal studies, T cells seemed to recognize and attack cancer cells. They latched on and released a poison, a protein called CD28 that should have destroyed the malignant cell, but the cancer somehow survived.

They learned that after the CD28 was released, another protein, CTLA-4, showed up. What was it doing there? A large pharmaceutical company had concluded it was another cell poison. That company patented it as a poison.

Allison wasn't so sure they were right, and he developed an antibody that blocked CTLA-4. One of his fellows gave it to a mouse with cancer. A few days later, the cancer was gone.

The results surprised Allison. Blocking a poison should not have contributed to the death of the tumor. Allison asked his fellow to repeat the experiment. Since it was Christmas, the fellow went on vacation and Allison manned the lab. He watched the mouse as the tumors grew for a few days, then faded away.

Allison immediately realized he might have something big, but he had to be sure. His group injected antibodies that blocked CTLA-4 into the bodies of many different strains of mice. In the absence of CTLA-4, the poison produced by the T lymphocyte—the CD28—was able to destroy one tumor after another.

Allison realized his success meant our understanding of cancer and the immune system was wrong. "The biology was backwards." T lymphocytes, he hypothesized, recognized cancer cells, and they latched on. They injected a "poison" (CD28) that should have killed the malignant cell. But cancer cells made an "antidote" (CTLA-4). It stopped the poison from working. His antibody blocked the antidote. It *allowed* the poison to keep killing the bad cells.

Bristol Myers Squibb had patented CTLA-4. Their patent claimed CTLA-4 was the poison, not the antidote. It was wrong—backwards. But the company had a patent and lawyers and money.

Allison was a valued scientist. His identification of the T cell antigen receptor was important. People in the field respected him. Bright, ambitious students studied with him. He was a full professor of immunology at Cal Berkeley. But he wasn't an M.D. His only interaction with sick people had occurred when, as a boy in a small Texas town, he had gone on house calls with his father, the town doctor. Allison wanted to try his antibody on patients, but he didn't think he could take the next step without pharma's help.

He "spent the following close to two years going around and talking to a number of large and a few small biotech firms. He tried to interest someone in his idea. There was a lot of skepticism. The fact that Bristol Myers Squibb had a patent put people off. They claimed the intellectual property was 'dirty.'"

Eventually, a small firm, Medarex, decided to give his antibody a shot. Niels Lonborg, a scientist at GenPharm, a company that was purchased by Medarex (in 1997), had mice that made fully human antibodies. Lonborg created the antibody to CTLA-4. It later became the drug ipilimumab.

Bristol Myers Squibb agreed to try the drug on patients. As Allison explained in an interview, he "was totally committed" to the endeavor. He moved to New York, to be near Sloan Kettering Cancer Center "to make sure nobody hurt his baby—Nobody screwed up." He moved to be a nuisance. The biology of ipilimumab (the drug's generic name) was different from that of most cancer drugs. "Usually you treat patients. If the tumor grows in the face of treatment the drug is a failure." But in the treated mice, the cancer grew for a while. Then it withered. The tumor didn't always regress, but "there was overall survival."

As shown in *Breakthrough*, a film that documented Allison's subsequent struggle, his discovery ran headlong into big pharma naysayers. Immunotherapy had failed in the past. To the conservative corporate money men, Allison's drug seemed like a long shot that was not worth the risk. The doctor coordinating the trials, Rachel Humphrey, believed in the product. She was its chief advocate. When she faced the pharma company board, she emphasized the fact that the drug had been effective in one person. That alone made it worth pursuing. Board members yelled at her and she took it.

A competitor, Pfizer, had an immunotherapy drug trial running at the time. They halted their effort when the tumors growing in their patients didn't shrink 30 percent in 12 weeks. That was the FDA standard.

With Allison's drug, the tumors kept enlarging but the patients felt well. Allison explained that's how the drug works. The T cell gets into the tumor and starts killing cancer cells, but it takes a long time before the tumor stops getting larger. As the drug trial progressed, some people who were treated went home and their doctor gave them a drug that had not previously worked. This time the old drug seemed to work. The person got well, and the old drug got the credit for the improvement. Allison knew they weren't responding to a drug that had failed and was being reintroduced. The people were better because the T cells had continued to methodically kill the tumor.

Bristol Myers Squibb, his company, eventually agreed the end point of their study would not be the number of people who were alive at one year or two years. They would instead see if there was an improvement in total survival.

At the end of three years, people who hadn't received immunotherapy were all dead. Three, four, and five years after they were treated over 20 percent of the people who received Allison's drug were alive and well. The company couldn't call it a cure. The tumor might someday start to grow. You never know. But the people who responded stayed well and the treatment sure acted like a cure.

Allison went to New York in 2004, but the drug wasn't approved by the FDA until 2011. During the intervening seven years, Allison lived in an apartment three blocks from the hospital. He gained weight and was frustrated by his need to keep explaining why the tumor was still present. At times he became angry. On one occasion, he went into a tirade. He'd come so far and he was afraid the company would conclude the drug failed. When he became single-minded and obsessed, it affected his relationship with his wife. Malinda, the woman he met when he was a college student, the coed who always felt he was the only person she ever loved—the most amazing human she ever saw—left.

In 1997, Medarex acquired GenPharm. In 2009, Bristol Myers Squibb paid Medarex $2.4 billion and the companies merged.

In the 1990s Tasuku Honjo, a prominent Japanese researcher discovered PD-1, a protein that keeps T lymphocytes from attacking cancerous cells. A researcher at Medarex developed the antibody that blocked PD-1 and allowed T cells to destroy certain types of cancer. In time additional drugs that block PD-1 and PD L-1 were fashioned, tested, and shown to prevent the growth of some malignancies.

In 2017 close to 1,300 people in twenty-one countries who had advanced melanoma, were assessed 3 years post treatment. All had received a combination of two drugs nivolumab (a PD-1 inhibitor) plus ipilimumab (a drug that targets CTLA-4). Fifty-eight percent were alive and in 39 percent the disease had not progressed. In another study, Dr. James Larkin of the Royal Marsden in England and others assessed three hundred–plus people who had metastatic melanoma. They had been treated with two drugs that blocked the antidotes produced by melanoma cell. "Historically, five-year survival rates among patients with metastatic melanoma were dismal," and the new treatment was hard on the bodies of the sufferers. Only 58 percent of the people treated seemed to have had a favorable response. But 52 percent of the three hundred–plus, most of who would have died without treatment were still alive at five years.

In 2018, the PD-1 inhibitor nivolumab "showed a clinically meaning-ful survival benefit in some people who had advanced lung cancer."

The drugs were the results of research performed at a public university and an antibody created by industry. Pharmaceutical companies paid for the costly clinical trials, the hospital days, evaluations, legal work, FDA submissions and advertising. After the drugs were approved for sale the manufacturers decided what they would charge.

Bristol-Myers Squibb listed the price of a course of treatment with ipilimumab (CTLA-4 Yervoy) at $150,000, and the price of a course of treatment with nivolumab (PD-1 Opdivo) at $150,000. The drugs brought in revenues of $7.5 billion in 2018.

The Merck PD-1 blocker pembrolizumab, Keytruda, fetched $7.1 bil-lion that year.

In 2018, Jim Allison shared the Nobel Prize with Tasuku Honjo, the investigator who discovered PD-1. When he received the award Honjo wore a traditional Japanese hakama and haori jacket, and he donated his half of the prize money to young researchers at Kyoto University.

The Cancer Treatment That Wants to Be a Commodity

T lymphocytes are capable of destroying cancerous cells, but they have trouble recognizing them. **B lymphocytes** are experts in identifying and marking targets, but they have no killer mechanism. What if the capabilities are combined? *Paraphrased words of Zelig Eshar, Weizmann Institute, Israel*

The researcher who envisioned inserting a seeing-eye gene into T cells was raised in Rehovot, Israel, at a time when "the fragrance of orange blossoms and the sounds of crowing roosters" filled the air. He earned a Ph.D. in Boston. While there, he "decoded" the T cell receptor. It's the molecule on the outer surface of the T lymphocytes that spots remnants of viruses that are displayed on the outside of sentry cells. During his 20 years at the Weizmann Institute in Israel, Eshar and his team developed and refined a gene that they planted in the cytoplasm of a T cell. It gave the cell the ability both to recognize and destroy worrisome proteins.

In another part of the world, a surgeon at the NIH, Steve Rosenberg, led a team that assembled a similar gene. Their CAR-T (chimeric antigen receptor) specifically targeted cells that have the protein CD 19 on their outer membranes. CD 19 is only found on the surface of B lymphocytes.

A compulsive researcher, Rosenberg once wrote that he enjoyed "working through the night in the lab, drinking thick pasty coffee that had been on the burner for hours, walking out into the sunrise, and watching the city come to life." During his forty NIH years, "there were probably only forty days when I wasn't in the hospital, checking on research or seeing patients." Over the years, he wrote a book about his efforts, was interviewed on TV by Charlie Rose, and was featured in Siddhartha Mukherjee's cancer documentary.

When he was young, Rosenberg was present when his father, a Jewish immigrant from Poland, received one postcard after another telling of relatives who had died in the death camps. The notifications evoked a depth of silence, and Rosenberg tasted pain. His desire to stop everyone's ache was not the sole reason he wanted to become a doctor, but it played a major role.

Steve was married to Alice, an emergency room nurse who disliked doctors' egos and was determined she would not marry one. The two met when he was on call in the emergency room. It was a slow night—not much business, and she called him over and led him outside to gaze at the moon-filled sky. They dated for five years before he told her, "We can't see each other anymore. Otherwise it will be too difficult to break it off." She answered, "It's already too late." After they married and had children, he was a surgical resident and worked many nights. He once fell asleep at a patient's bedside and he routinely dozed when he stopped for a red light when driving home. His sleep deprivation was brutal.

In 1968, as surgical resident, he recalls admitting a man with a gall-bladder attack who had undergone a cancer operation twelve years earlier. The primary malignancy was in the stomach. When it was cut out, there were metastatic implants in the liver that could not be resected. The man should have died, but he didn't. His immune system apparently overcame the cancer. It happens rarely, but the event made an impression on the young doctor.

In 1974, Rosenberg started working at the National Cancer Institute. He began his search for a way to help the body's immune defenses fight cancer.

In 2010, he and his colleague James Kochenderfer told their medical colleagues about a patient with lymphoma whose tumors shrank after they altered his T cells. They added a gene that allowed his T lymphocytes to recognize CD 19. That's an antigen that is only found on the surface of the healthy and malignant B cells. After the gene was added, they allowed the changed T cells to multiply. Then they poured them back into the person's bloodstream. The patient became and remained cancer free. Rosenberg subsequently treated an occasional patient using the same approach. He also sent a copy of the gene in a plasmid to Addgene.

Addgene is a nonprofit that's based in Cambridge, Massachusetts. Since it was founded in 2004, the company has collected over 80,000 gene-containing plasmids from over four thousand research labs around the world. When researchers have isolated a gene, performed research, and written a paper, they often want to move on. If they want to make their gene available to others, they can package the gene in a plasmid and send it to Addgene. The company preserves and catalogs the important chains of DNA and they announce their presence on their website. In 2010, researchers could buy one of the plasmids that contains a gene for $75.

In 2011 Dr. June, at the University of Pennsylvania, wrote about two of three patients with chronic lymphocytic leukemia who were similarly treated and went into complete remission. At the time, he didn't think the infusion was ready for general use. As Dr. June explained, "Some of these responses don't last—there's resistance. We still have to run rigorous randomized studies to determine if the therapies are effective, and whether they are cost-effective, and whether they can be delivered at scale."

CAR-T infusion often takes a month to prepare. It's a therapy some doctors consider when they are dealing with someone with a difficult-to-treat lymphoma. Once the patient is identified and consents, a large-bore needle is inserted into a vein. It is hooked to an apheresis machine. Blood is drawn into the sophisticated instrument. A centrifuge spins the blood, picks out the T cells, and returns the rest of the blood to the patient. The collection is then "prepared, frozen, and sent to the facility where CAR-T genes have been inserted into a number of harmless viruses." The viruses are allowed to infect the T cells. They deposit the gene into its cytoplasm, and the T cells are given time to reproduce and increase in number. Before they are infused into the person with lymphoma, the patient is sometimes given intensive chemotherapy. After the altered T cells are infused, the patient is watched carefully for up to thirty-five days. The death of a large numbers of tumor cells can cause the body to release cytokines and they can make the patient very sick.

In the years after the process was developed, a few individuals who had no other option were successfully treated. The approach was new, expensive, and time consuming. There were risks. The short-term improvements may or may not have meant the person's life would be prolonged.

In 2009, a UCLA urologist and businessman, Arie Belldegrun, founded Kite Pharmaceuticals in Santa Monica. He searched the academic market to see who, if anyone, knew how to use the immune system to fight cancer. Years before, when he was a young doctor, he had briefly worked with Steve Rosenberg on cancer immunology. He eventually contacted his old boss. Dr. Rosenberg showed him the X-rays of several patients he had successfully treated with CAR-T. It's a onetime treatment. When gene therapy works, the cancer is gone in three to four weeks. Rosenberg wanted to commercialize the process and tried to interest Johnson & Johnson and a few other companies in the project. The companies all thought the approach was too new and different.

Belldegrun was impressed. "In 2012, Kite Pharma partnered with Dr. Rosenberg and the NCI (National Cancer Institute) to further the research. They are attempting to develop multiple chimeric antigen receptors (CAR) and T cell receptor (TCR) based products."

In 2018, the FDA gave two companies permission to sell a new, unproven type of immunotherapy that treats lymphoma. In one study, CAR-T cells were used in kids. They eliminated malignant cells 83 percent of the time for at least three months.

To get FDA approval, Kite had to prove their approach was effective. Since it's unethical to run a study where some qualified patients are not treated—where there are "controls"—the FDA is allowing the company to use historical controls to prove their approach is effective. By the end of February 2020, 108 people had been followed long enough. Kite was finally able to say they had proven the approach prolongs the life of at least some people.

Kite, Novartis, and other companies are starting to offer CAR-T treatment. Kite charges $375,000 for processing a person's lymphocytes. In 2017, Gilead purchased Kite for $11.9 billion. Three years later, Medicare said they would pay for the process.

What about Canadian Pharmacies?

Under the Prescription Drug Marketing Act of 1987, it is illegal for foreign "e-pharmacies" or anyone other than the original manufacturer to bring prescription drugs into the country. I've used a Canadian pharmacy to buy the eye drops I needed for my glaucoma. After choosing an online pharmacy that, best I could tell, was legit, I placed my order, supplied proof that I was a licensed physician, and emailed a handwritten prescription.

My prescription was processed in Canada and the medication was mailed directly to my home. It did not go through Canada. It was not sent to me by a Canadian pharmacist. The medication came from a factory in Germany (once) and Turkey (once). Manufacturers legally ship medications to the U.S. all the time. My medicine was legally brought into the country.

In most states, drugs ordered through Canadian pharmacies are not paid for by insurance or Medicaid or Medicare. They require a lot of effort by doctors and patients, and they take two weeks to get processed. If the drugs were recently approved by the FDA, they often aren't cheap or available.

Half of the businesses that sell pharmaceuticals on the web are located in the U.S. Some call themselves prescription referral services. The National Association of Boards of Pharmacy claims, "Rogue websites may be selling drugs that are counterfeit, contaminated, or otherwise unsafe." I'm sure they are right. People have always been tricked by charlatans promoting cure-alls. It's hard to know how often we or our friends have been fooled by Internet fraudsters. People who use Canadian pharmacies need to be attentive and cautious. There's a web-based company that claims it earns its keep by certifying their legitimacy. Called PharmacyChecker.com, the "verifying" company charges pharmacies an annual fee. Each medication filled must have a valid prescription. A licensed pharmacist has to make sure the "medication is selected and labeled correctly." In countries with the strongest regulations, Pharmacy Checker inspectors don't visit pharmacies; the company merely verifies the pharmacy's license.

CHAPTER TWENTY-EIGHT

Gene Therapy and CRISPR

E ach year, one hundred American babies are born with Pompe's disease. The children are floppy, their muscles barely work, and their heart is enlarged. Few survive infancy. A genetic, recessive condition, the disease is only seen when both parents carry the defective gene. The malady is the result of an enzyme deficiency. The kids' cells don't make enough lysosomal acid alpha-glucosidase. It is a protein that's used to convert stored glycogen into glucose—energy.

We eat carbohydrates and sugar. We turn what we don't use into a storage polysaccharide called glycogen. Then we stockpile the excess fuel in our muscles and liver. Between meals, when our bodies need sugar, we turn glycogen back into glucose. Babies who lack the enzyme can't break glycogen down. As the glycogen accumulates, it progressively weakens the muscles of their heart and body. Kids with the condition don't live long.

The condition was "characterized" by and named for a Dutch pathologist named Joannes Pompe. A member of the Dutch resistance, he was executed by the Nazis in 1945. The absent enzyme was isolated in Belgium in 1955, and the responsible gene was identified in 1979. (It differs a little from one family to another.)

The enzyme the babies lack was first made at Duke University by a dedicated team of researchers. Their leader, Dr. Chen, the chief pediatrician, started his quest after he went to the funeral of an infant who died of the disease. The pastor said God must have given the child life for some reason. Chen took the message to heart, and decided to assemble a team of Duke University researchers and go to work. He began his research in 1991, the year a team of researchers from NYU isolated and determined the DNA sequence of the needed GAA gene. I don't know how Chen's group got hold of it. Duke researchers apparently inserted the gene into plasmids and inserted the plasmids into cells derived from the ovaries of Chinese hamsters. It is a mammalian cell line that is frequently used for mass production of therapeutic proteins.

It took the researchers three years to make enough of the needed enzyme for their early tests. Once they had sufficient juice, they injected it into a Japanese quail that had a glycogen storage disease. Before it was treated, the poor bird couldn't get off its back, much less fly. Post injection, the creature stood and even flew a little.

After six years of successful research, the Duke scientists got some manufacturing help. Production rights were licensed to Synpac, a British/Taiwanese company with a presence in Durham, Duke's home. Synpac, in turn, "used experienced contractors to manufacture the enzyme." Having done the heavy lifting, the Duke scientists gave a lot away, but they retained some royalty rights.

Once the company had produced enough, six kids who were in bad shape were infused with the enzyme. They received a dose every one to two weeks for a year. At the end of that time, all diseased heart muscles were pumping better and all the kids were growing. Three children were able to walk, "an achievement unlikely to occur in patients who are not treated." Enzyme replacement worked.

In 2006, Synpac and Genzyme signed a "fifteen-year royalty-sharing agreement that was potentially worth $821 million." In 2010, Genzyme was based in Boston, and had $4 billion of revenue. The company planned to spend more than $500 million to create production facilities for Myozyme (their name for the enzyme). But Myozyme had become a valuable commodity, and the following year, Genzyme was acquired by the French pharmaceutical giant Sanofi for $20 billion. As part of the process, Duke University was paid $90 million, and the university relinquished its royalty rights. In 2016, Sanofi sold $800 million worth of the needed enzyme. Now called Lumizyme, the enzyme is currently made in large sterile factories. Babies with the disorder get an injection of the protein every two weeks.

In this country, "according to Sanofi, a year of therapy, on average, costs $298,000." It's not curative. If it is effective, the treatment has to be repeated each year. By the time a child is ten years old—if no one develops a less expensive generic product—the system (insurance companies and Medicaid) will have shelled out $1 to $3 million dollars per child, and big pharma will have been handsomely compensated.

Insurers, of course, are not allowed to deny coverage to a person who has a preexisting condition. They can't impose lifetime or annual coverage units. But they are allowed to turn a 15 to 20 percent profit. If their earnings go below that level, insurers can charge more for their policies.

The first marker for a genetic disease was found in 1983. It identified Huntington's, a dominant malady that strikes in midlife. The disorder became well known after Woody Guthrie learned that his jerky movements, rigidity, clumsiness, and inability to think clearly were caused by the condition that had destroyed his mother. Guthrie is the Oklahoma folk singer who wrote "This Land Is Your Land." He was forty years old when the symptoms started.

When he was fourteen, Woody was "abandoned." His mother was hospitalized with Huntington's and his father had to move to a nearby town for a job. The family was quite poor and the country was in the midst of a depression. Guthrie spent some of his teenage years sleeping at various friends' homes and rejoined his dad after a few years. When he was a teenager, Guthrie was more interested in his guitar than he was in high school. At age 19, he married, and the couple had three children. During the 1930s, he and his family were living near the Oklahoma panhandle. Giant clouds of dust periodically blew in. The swirls of dirt filled lungs and killed cattle and a few children. The area was barely habitable. When others pulled up stakes and headed to California, Woody decided to join them. Two of Woody's first three children developed Huntington's in their early forties. During his fifty-five years, Woody served in the merchant marines, lived in California and New York, and was a popular entertainer. He married two more times and wrote one thousand songs. One of them turned out to be his farewell message: "So Long, It's Been Good to Know Yuh." The abnormal gene that is responsible for Huntington's has been isolated and studied. There is currently no treatment for the condition.

The vast expanse of DNA in the nucleus of each cell contains the twenty thousand genes that are unique to each person. The genes account for about 3 percent of the nuclear DNA. Over several decades, researchers identified the nucleotide sequences responsible for one genetic disease, then another. In 1990, scientists started mapping the entire human genome. The task was "declared completed" in April 1993.

Every so often, a child is born with one of over six thousand genetic disorders, and a few families have to raise a disabled infant who will die young. Some researchers working on the problems have developed treatments that supply or teach the body to replace a vital protein. After a protein is made, it has to coil and fold into a specific three-dimensional shape. Misfolding produces inactive or toxic proteins and is the cause of a number of genetic diseases.

There was a time when it was hard to get FDA approval for diseases that only affected a small number of people. The agency required large trials before they would conclude a new drug was safe and effective. Then parents got together and pressured members of Congress. In 1983, legislators passed and the president signed the Orphan Drug Act. It gave a lot of rewards to companies that manufactured drugs for fewer than 200,000 Americans. Their FDA monopoly lasted seven, not five, years. The companies got tax credits and were able to write off half of the development costs. If the disease was rare, developers skipped the usual wait and joined the "fast-track" line.

There are seven thousand rare diseases affecting 25 million to 30 million Americans. In the first twenty years after the law was passed, 249 orphan drugs were marketed. It's estimated that by 2024 "orphan drugs will make up one-fifth of worldwide prescription sales and bring in $242 billion. In 2018 the cost per patient per year of the top one hundred orphan products was $150,854." Insurance companies that stay in the market and ultimately the taxpayers will soon have a new flood of costs they will increasingly have to deal with.

Cystic fibrosis is genetic, and recessive. If both parents are carriers, one of four offspring is afflicted. In kids who have the condition, the lung mucous that collects bacteria and foreign particles is *not* watery. It's not easily swept out of the lungs and swallowed or coughed up. It's thick and "glue like." People with the disease have a hard time getting rid of it. Kids periodically develop pneumonia. Over time, they lose lung function. A century ago, most of the afflicted weren't aggressively treated with inhaled bronchodilators, physical therapy, postural drainage, and appropriate antibiotics. Few survived childhood.

The Cystic Fibrosis Foundation has established 117 centers of excellence. They are manned by experienced health care professionals and have guidelines, "best" practices, and public monitoring. As a result of their aggressive approach, the average person with cystic fibrosis now lives an average of thirty-five years, though getting kids through their teenage years is typically tough.

In recent decades, a few medications that make a difference were developed by Vertex. It's a biochemical start-up that spent $4 billion during its first twenty-two years without developing an approved drug. The company was initially reluctant to get into the cystic fibrosis business. There were only thirty thousand Americans with the condition. If researchers found a potential drug, it would cost $100,000 to test each person. In the companies' minds, the expense of becoming involved was "prohibitive." Richard Aldrich, a deal maker and advisor, thought Vertex should only work with the CF (Cystic Fibrosis) Foundation "if the foundation agreed to fund some of the early stage (drug) development."

The team that led the company's effort was headed by Eric Olson, an experienced research biologist. He was interested in CF in part because a colleague/friend's daughter had the disease.

The faulty piece of DNA that causes the condition was located in 1989 by a group of Canadian geneticists working with Hong Kong–born

Lap-Chee Tsui. The mutation responsible for most cases of cystic fibrosis occurs when three nucleotides are deleted from a gene on chromosome 7. Called cystic fibrosis trans-membrane conductance regulator (CFTR), the abnormal gene causes the cell to make a membrane protein that doesn't "fold" normally. Appropriately folded protein regulates the amount of chloride, salt, and water that flows in and out cells. The fluid travels through "channels" in the cell's outer wall or membrane. In people with cystic fibrosis, secretions are thick, and salty sweat accumulates outside the cell.

Vertex researchers were funded by $47 million from the Cystic Fibrosis Foundation and $20 million from the Gates Foundation. By June of 2011, they had developed two drugs that dropped the salt content of sweat. The medications reduced the exacerbation rate. They decreased the unexplained "worsenings" that contributed to a more rapid decline in pulmonary function. In the early 2000s, Vertex added a third drug. It had an effect on people who have a "Phe508del CFTR mutation." It's the most common abnormal gene responsible for the disease. Ninety percent of people with cystic fibrosis have at least one copy of the mutated DNA. When it was given to kids, it improved lung function and sweat chloride. It also lowered the number of "pulmonary exacerbations"—worsening that sometimes led to hospitalizations.

One analyst felt the triple-drug combo will rake in close to $4.3 billion by 2024. In August 2020, the executive branch of the European Union "approved" the new treatment for use in the ten thousand Europeans who are twelve years and older and have the appropriate genetic mutations.

The most common genetic cause of death in infancy, spinal muscular atrophy, "causes severe weakness by six months of age and inability to breathe by the age of two." It occurs when the gene that directs cells to make an essential protein is deleted or mutated. In the absence of the protein, the nerves that send signals to muscles die.

People are born with a second gene that's similar and that isn't genetically affected. It's not able to make the needed protein, and in healthy people that's OK. Most of us don't need it.

Scientists at Cold Spring Harbor Laboratory, a large nonprofit research center on the north shore of Long Island, developed a "segment of RNA." When it is injected into the spinal fluid, it allows the second gene to make the needed protein. It was a great accomplishment and it wasn't easy. When the drug was approved by the FDA, the company that owned it decided to charge $125,000 for each dose, or $750,000 the first year and half as much each subsequent year. A physician at the University of Utah cares for "about 150 patients with the mutation. He complained in an article that if each child he cared for was treated with nusinersen (the segment of RNA), the cost would be $113 million the first year and $56 million annually thereafter."

In November 2017, an Ohio company developed an alternative approach to the problem. Their therapy was based on research performed by Brian Kasper at Nationwide, a Columbus, Ohio, children's hospital. An employed researcher, Kasper was studying adeno-associated viruses (AAVs). The organisms don't cause significant disease and are sometimes used to carry genes into the cytoplasm of a cell and deposit them. One day a member of Kasper's team discovered a viral serotype that penetrated the blood-brain barrier. That was unusual. There are fifty serotypes of adenoviruses that don't make people sick. Most can't get into the brain.

Kasper and his team believed the virus they discovered provided "a new way of delivering genes to widespread regions of the central nervous system." The drug companies they approached allegedly weren't interested. Then, in 2013, with the help of a biotech entrepreneur, Kasper formed a start-up, AveXis. They raised $75 million and licensed the therapy from the Columbus hospital. To this point, all research and development was paid for by the U.S. government and charitable funds.

The company researchers placed a gene that promoted the production of the needed protein into an adenovirus. Then they infused a high dose of the virus that contained the gene into the bloodstream of twelve affected children who were about six months old, and the genetic therapy made a major difference.

After one-and-a-half to two years, eleven of the children were able to speak. Nine could sit unassisted for at least thirty seconds, eleven achieved head control, nine could roll over, and two were able to crawl, stand independently, and walk independently.

The start-up spent some of its $75 million and probably purchased the genes from Addgene. In April 2018, Novartis bought AveXis for $8.7 billion. After the FDA approved the therapeutic approach, Novartis named the treatment Zolgensma and decided to charge $2.125 million for each child they treat.

Currently "there are more than eight hundred cell- and gene-therapy programs in clinical development. Several of the medications have been approved by the FDA," and gene therapy is in its infancy.

Some of the remedies on the market are owned by Biomarin, a company headquartered in San Rafael, California. Founded in 1997, the company has acquired six biomedical start-ups in the last fifteen years, and in 2016 was marketing five orphan drugs.

Genetic engineering has shown promise in people with hemophilia. The genetic condition afflicted the royals of Europe and played a role in the 1917 fall of the Russian Tsar. It occurs in one in five thousand live births, and it is sex linked. That means women carry the gene and their sons get the disease. When an affected male is injured and his factor eight or nine is low, he can't stop bleeding. Much as people are unable to climb up a ladder that is missing a rung, men with hemophilia are missing a protein in the clot-forming ladder and can't develop a thrombus. When their levels are low, their joints fill up with blood and over time they develop deformities. The joint complication is very painful. When the blood level of clotting factor number eight is at or below 1 percent of normal, the condition is "severe." When a person's blood has 5 to 40 percent of the factor, they have a moderate problem and they mainly receive factor infusions before surgery or if there is a need.

Victoria, the queen of Great Britain from 1837 to 1901, was a carrier of the hemophilia B gene. She passed the condition through her daughter Alexandra to her grandson Alexei, the only son of Russian Tsar Nicholas. At one point, the boy's painful and frightening bleeds seemed to be helped by a self-proclaimed holy man named Rasputin.

During the First World War, Tsar Nicholas spent a lot of time at the front, and his wife was in charge of the government. She was born a German, and Russia was at war with Germany. She also seemed to be "under the spell" of the holy man, Rasputin. The war went badly for

Russia. Over 5 million soldiers were killed or wounded and the Russian people rebelled. In March 1917, the Tsar was forced to abdicate. Many believe hemophilia played a major role in the fall of the Romanoff dynasty.

During the height of the AIDS epidemic, blood banks used hundreds of units of blood to gather and pool the factor that stopped hemophiliacs from bleeding. Some of the blood they used came from people who had HIV and didn't know it. At the time, HIV was a lethal condition.

Researchers at the University College London recently put part of the factor 8 gene into an adeno-associated virus and "infected" a number of men. It caused their liver to make the missing protein. In six of seven patients who received a high dose of genes, "factor 8 increased to a normal level and stayed there for a year. None of the seven bled during that time. After two to three years the treatment was still providing a clinically relevant benefit."

An estimated one hundred thousand Americans have sickle cell disease. It's a genetic abnormality that affects adult hemoglobin but doesn't influence fetal hemoglobin. Hemoglobin is the protein that fills red cells. It allows the cells to pick up oxygen in the lungs and deliver it to all parts of the body. Affected people make red cells that aren't round and flexible. They look like sickles or crescent moons, and they clump, stick in small blood vessels. By so doing, they cause severe pain, anemia, stroke, pulmonary hypertension, organ failure, and early death. As explained on *60 Minutes*, Francis Collins of the NIH thinks we can cure sickle cell anemia. The approach he discussed uses CRISPR gene editing to increase blood levels of fetal hemoglobin (HbF). Hemoglobin F is the form of hemoglobin that fetuses use to extract oxygen from the placenta and deliver it to their bodies. Shortly after birth, a gene causes most children to stop producing Hemoglobin F, and adult hemoglobin takes over. Researchers at Vertex and CRISPR Therapeutics collected stem cells from a person with severe sickle cell disease. In the lab, they used CRISPR to destroy the stem cell gene that shuts down production of fetal hemoglobin. They then used chemotherapy to destroy the person's normal bone marrow. Finally, they infused the edited cells into the patient. Once the red cells of the person were filled with fetal hemoglobin they didn't "sickle." The therapy seems to be working.

CRISPR-derived gene therapy is new, exciting, and it will be widely used in future gene editing. The investigators who did many of the studies and developed the concept were publicly funded.

They were initially just trying to learn how bacteria defend themselves from assaulting viruses. They were also trying to make sense of an unusual stretch of bacterial DNA.

The following theory of how CRISPR came into being helped me understand the process: When a virus assaults a bacterium, the invader enters the cell, takes over its DNA, and directs the bacteria to make billions of viral particles.

Most bacteria are enslaved, then destroyed. A few mount a defense and survive. Some of the survivors create a "DNA memory file." They keep it in the CRISPR stretch of the bacteria's DNA. It contains the identifying characteristics of the harmful virus. In subsequent generations, the memory DNA creates strands of RNA that float around inside the bacteria. When a segment of RNA recognizes an invading virus, it latches on and cuts the virus apart with an enzyme called Cas9.

The group that figured out how to use the system to edit genes was led by University of California professor Jennifer Doudna and Emmanuelle Charpentier. Doudna was the daughter of a professor of English. She grew up in Hawaii and spent the summer that followed her college freshman year in a lab studying a fungus that was invading papayas. "It turned out to be a lot of fun," she hungered for more, and she worked in a few labs and made a few discoveries. After a decade, she became the head of a research lab at the University of California in Berkeley. Its campus was often blanketed in fog. On clear days it provides a spectacular view of the Golden Gate Bridge and the Pacific beyond.

Doudna met fellow researcher Emmanuelle Charpentier at a conference in Puerto Rico. The two women discussed possibly collaborating while they explored the narrow, cobbled lanes of Old San Juan. Charpentier was "small and slight, with eyes so dark they almost seemed black." When she was a Ph.D. student at the Pasteur Institute in Paris, Charpentier "realized she had found her environment"—she was hooked on the life. After Paris, she spent more than twenty years performing research in nine different institutes in five different countries. Years later, reflecting on her life, Charpentier commented that becoming a scientist was "a little bit like entering a monastery." She worked late, was rarely in bed before midnight, and had "a very bad tendency to wake up in the middle of the night and work."

She had come to Puerto Rico and sought a meeting with Doudna because she discovered an RNA molecule that helps guide Cas9, the enzyme that cuts DNA.

In June 2012, Doudna and Charpentier published a study that showed how RNA and Cas9 could be used for "site-specific DNA cleavage and RNA-programmable genome editing" (cutting out the old defective gene and replacing it with a new flawless one). Investigators around the world took notice and got busy.

After they understood how bacteria identified and destroyed unwanted viruses, Jennifer Doudna and other researchers tried to use the system to edit genes. They chose a target—a "twenty-letter DNA sequence" that was part of the gene they wanted to delete. They then rearranged a sequence of RNA building blocks (nucleotides) and created a matching twenty-letter strand of RNA, The RNA was used to guide the Cas9 past the cell's 3 billion pairs of DNA nucleotides. It identified and "coupled" with the desired strand of DNA, and the Cas9 knife cut through the DNA.

The researchers planted the genetic instructions for making Cas9—the knife—into one plasmid, and they put the genetic instructions for "guide RNA" into a second plasmid.

Their concoction was able to search a cell's DNA—3 billion pairs of nucleotides—find and unwind the desired segment of DNA--and use Cas9 to cut apart the strand of nucleotides. In other words, they could irreparably damage a chosen gene.

Cells know how to repair a break in their DNA. The cut ends either come together on their own, or the gap can be bridged by a segment of DNA—a gene.

Scientists already knew how to add a new, good gene. Years earlier, Mario Capecchi had learned that genes intuitively know where, amidst the 3 billion pairs of DNA nucleotides, they belong. If good genes are developed and put into cells, they find and attach to "their place."

After Doudna's paper was published, Kevin Esvelt (currently at MIT) explained how using CRISPR and selfish germ line genes can create changes that will be inherited by future generations of cells.

Sensing that there wasn't time to write grants and get government funding, Doudna and other scientists formed a venture capital company—Editas Medicine. Charpentier and others founded CRISPR Therapeutics. Both firms, according to their websites, are trying to cure sickle cell disease, cystic fibrosis, and a few other genetic conditions.

In May 2012, a month before Doudna and Charpentier published their CRISPR paper, the University of California filed a patent claim for the intellectual property surrounding the process. While Doudna and Charpentier were hard at work in California, researchers at the Broad Institute of MIT and Harvard were also working on CRISPR. They independently proved it could be used as a gene editing tool. In December of 2012, they also filed a patent for CRISPR. The scientific paper that showed their accomplishment was published the following month. Then Broad paid the U.S. patent office to fast-track their claim. In April 2014, the Boston group was granted the first patent. By now a lot of venture capitalists were involved and they knew CRISPR was a valuable commodity. The lawyers got involved and are still at it.

In the fall of 2020, Doudna and Charpentier were awarded the Nobel Prize in chemistry.

CHAPTER TWENTY-NINE

Gaming the System

"To be audacious with tact you must know when you've gone too far."
—JEAN COCTEAU

When I started learning about Martin Shkreli, I expected to find a canary in a coal mine—the Edward Snowden of drug prices—an in-your-face rebel trying to force the country to stare hard at its absurd drug pricing system. But that wasn't who he was. Shkreli was the son of immigrants. He went to business school and worked on Wall Street and allegedly (when young and new) exaggerated or lied to some of his hedge fund clients. He insists the people who stuck with him all made a profit, so no harm done. Later, he became the head of a pharmaceutical company that overpaid for a drug called Daraprim.

It was one of many drugs developed by Gertrude Elion, the daughter of a Lithuanian dentist who was bankrupted by the 1929 stock market crash. An accomplished student, Elion went to Hunter College tuition free and graduated near the top of her class. When she sought work in a lab, she learned there were no jobs for women. The next few years, she lived at home and went to graduate school at night. She worked odd jobs like teaching part-time, checking the acidity of pickles and the color of mayonnaise for a grocery chain, and testing the strength of sutures for Johnson & Johnson. When she was finally able to study advanced chemistry, she noticed she was the only female in the class. During the Second World War, men were in short supply and she was hired as an assistant to a pharmaceutical researcher. She spent thirty nine years at Burroughs Wellcome making molecules that antagonized the action of purines, chemical structures that are components of DNA. Some of the drugs her group developed helped doctors fight some leukemias. Others modified a body's immune system. In 1988, she won the Nobel Prize. Daraprim, one of many medications she created, was owned by GlaxoSmithKline and was initially used for malaria. Doctors stopped using it after many of the parasites became resistant to the medication. By the time Shkreli took control, Daraprim had been around for decades and it couldn't have been very profitable. It was sometimes used to treat toxoplasmosis. The parasite is transmitted by cats and can damage the eyes and brains of newborns. People with advanced AIDS have immune systems that have lost their ability to protect a person from the microscopic organisms that once lived harmlessly in their bodies. When someone had advanced AIDS, toxoplasmosis could invade the central nervous system. After it got a foothold, the parasite could cause focal cranial lesions and encephalitis, a potentially lethal inflammation. The treatment of choice for a toxoplasmosis brain infection was a combination of Daraprim and sulfadiazine.

A little over a million Americans are living with HIV. The majority are taking drugs every day. Their virus is suppressed and is not destructive. Fifteen percent of those infected are unaware. A little over six thousand annual U.S. deaths are attributed to HIV. I don't know how often toxoplasmosis contributes to their demise.

Made in a few places in the world, Daraprim had long been available and cheap. GlaxoSmithKline couldn't or wouldn't raise the drug's price for practical, philosophic, and public perception reasons. So it was kind of a financial loser and was one of several drugs that was sold or dumped by GSK. Daraprim was briefly owned by a drug company called Tower Holdings. They somehow turned it into a commodity, something of value they could trade. It was passed along, and after a series of drug company acquisitions and mergers, the medication became the property of Impax of Hayward, California.

The company (allegedly) claimed they sold $9 million worth of the medication each year and made little or no profit. In August 2015, they convinced Turing Pharmaceutical to buy the drug for $55 million. At the time, Turing was a privately held start-up with offices in Switzerland and New York. According to its LinkedIn page, the company once had fifty to two hundred employees.

After the drug was acquired, Turing tried to start a campaign to make mothers aware of the possibility of transmitting toxoplasmosis to their fetus. I'm not sure what they were thinking. Toxoplasmosis in newborns is uncommon. Of the 4 million children born in the U.S. each year, an estimated four hundred have toxoplasmosis. That's .01 percent. When Daraprim (pyrimethamine) is fed to pregnant animals, many of their off-spring are born with abnormalities. So we avoid giving the medication to pregnant women.

Once Daraprim was part of Turing's arsenal, the company's CEO, Shkreli, raised its price from $13.50 to $750 per pill. A self-professed Republican, he chose not to explain the price hike. It was legal. Drug companies raise their prices all the time.

According to Barry Werth, author of *The Billion-Dollar Molecule*, "You can get away with high drug prices if you do it right. If he had raised the price thirty times instead of five thousand times, he could have gotten away with it."

Because of its high cost, pharmacies and hospitals were reluctant to stock the medication, and it was hard to obtain on short notice. Then a person with AIDS was hospitalized with a toxoplasmosis brain infection and his doctor found it difficult to get the drug. Dr. Judith Aberg, the head of infectious disease at a New York medical center, got involved and was outraged by the price. She told the person's story to the *New York Times* and Shkreli was vilified.

When he appeared before a congressional committee, he wasn't con-trite, refused to answer questions, and took the Fifth. The amount charged for the drug was legal, and Shkreli was not apologetic. He was interviewed repeatedly by journalists and became infamous.

His company, Turing, was sued by Impax. They no longer owned the drug but, reportedly, still owed the government $30 million for Daraprim-related Medicaid requirements. Impax wanted Turing to pay, but a judge found it wasn't part of the contract. Then the Department of Justice indicted Shkreli for alleged misconduct as a hedge fund manager. He was found guilty on five of eight charges, he lost his job, and the company laid off a lot of people.

As Shkreli said on an Internet talk show (edited), "In life you can play the game or you can give up the fakeness and be yourself. We saw a major insider trading case that was settled by the SEC. Big banks take millions in fines. No one gets arrested. There was a security charge that I manipulated stock price and another that I defrauded investors. They all made money."

"People with insurance or under Medicaid don't pay for their drugs. They pay co-payments." Much of the cost of Daraprim, like most expensive drugs, is borne by the taxpayer or it becomes part of the rising cost of health insurance. And that wasn't Shkreli's problem. He wasn't a rebel. He had no cause.

I don't know what Daraprim costs today. Shkreli, who was out on bail, was rearrested for a stupid Internet prank and the judge had him imprisoned. I have no idea what makes him tick. Numerous Americans are bothered by the cost of the medications they take, but they don't usually focus their ire on the big pharmaceutical companies. Instead, they talk about "the gougers"—the notorious few who dramatically jacked up the price of the drug that helped a few people who were dying of AIDS.

A few years back, a Philadelphia drug manufacturer got exclusive FDA rights (a many-year U.S. monopoly) to a drug I had been using for forty years. The medication, colchicine, is a plant extract that was used to treat gout before Jesus was born. It is one of a handful of ancient cures that withstood the test of time. The flower that produces the alkaloid was introduced to the New World by none other than Ben Franklin, an innovative guy who used the ancient remedy to treat his painful joints. I learned about the medication in medical school, and have advised many to take it. It has its share of side effects, and over the years has helped many of my patients, while making a few sick. The books back then told doctors to give repeated doses to people with acute painful swelling of the joint of the big toe—gout. We weren't supposed to stop until the pain subsided or the patient became nauseated or developed loose bowels. That turned out to be too aggressive for a few of my patients, and I quickly adjusted my approach.

For centuries, physicians have successfully used it, but no one did a double blind controlled study. Most docs would have thought withholding the drug from the control group would neither be necessary nor ethical. Then, thirty-four years ago, doctors in New Zealand did the study. Their 1987 paper was titled "Does colchicine work? The results of the first controlled study in acute gout." Half the people with an acutely inflamed joint took the real drug, and the other half took a placebo (an inert look-alike pill). People taking colchicine improved more rapidly and more completely.

In the company's defense, clinical experience is sometimes misleading. On occasion, useful drugs fail or people get well in spite of us. But colchicine has been used a lot over the centuries, and if the test of time means anything, the medication has always passed with flying colors.

The drug was available, cheap, and on the pharmacy shelf. No one had to go to the FDA to bring it to market. Then some whiz kid figured out how his company could get exclusive rights to the old herbal remedy. They ran a trial where neither the investigator nor the patient knew what substance was being used. (Though, frankly, it's hard not to know when the pill you are testing causes nausea and diarrhea at high doses.) Colchicine, of course, worked. The results were presented to the FDA and the whiz kid's company got exclusive rights to sell the herb extract in this country. "After the FDA approved Colcrys, the manufacturer brought a lawsuit seeking to remove any other versions of colchicine from the market. They raised the price by a factor of more than fifty, from $0.09 per pill to $4.85 per pill." Since this was a widely used medication, they apparently stood to take in an additional $50 million a year during the next seven years. The manufacturer received three years of exclusivity for gout and seven years for Familial Mediterranean Fever (FMF), although no new FMF studies were conducted. Outside the U.S., colchicine still costs $0.09 a pill. (That's how the letter of the law, and common sense, sometimes work in this country.)

On May 6, 2018, the TV show *60 Minutes* explored Mallinckrodt Pharmaceuticals' decision to sell Acthar Gel for $40,000 a vial. It had been priced at $40 seven years earlier.

Acthar is the brand name for a hormone called ACTH. It is one of many hormones made in the pituitary, the small endocrine gland located at the bottom of the brain. It is extracted from slaughtered pigs, and it directs the adrenal gland to make cortisol. In the early 1950s, it was our only means of giving some patients the hormone, but in 1955 prednisone became available and doctors largely stopped using Acthar. The product then had but two "accepted" indications: It uniquely helped a rare seizure disorder known as infantile spasm, and it helped doctors determine the reason the adrenal gland was failing to make the hormone cortisone.

By 2001, doctors were only prescribing Acthar now and then, and it was a money loser, but some kids needed it. Its manufacturer, Aventis, apparently felt someone should keep producing it. The French company managed to sell the drug to Questcor, a California "pharmaceutical company" that was losing money. Questcor paid $100,000 for the commodity, raised the price, and promoted the hormone for a few additional "indications."

The drug was subsequently administered by some doctors in place of other cortisone derivatives like prednisone. It could be used to treat a joint or kidney inflammation and some multiple sclerosis flares. Like prednisone it had side effects and created a sense of well-being. The act of getting a shot in a doctor's office seems to make some people feel better and Acthar injections are probably good for business.

The Acthar promotional campaign worked, and Acthar became a drug of value. In 2013, *Forbes* named Questcor the year's best small company. In 2014, Mallinckrodt paid $5.6 billion for the business and its moneymaker, Acthar.

In 2018, the FDC charged Mallinckrodt with price fixing. To keep the price high, the company paid Novartis $135 million and acquired the rights to Synacthen, a drug that is biologically similar to Acthar and was selling for $33 in Canada. Once they owned the alternative drug, Mallinckrodt "put their drug's competitor on the shelf."

Mallinckrodt was charged with antitrust. Admitting no wrong, they settled the case for a hundred million dollars. The company makes more than a billion dollars a year on Acthar alone. With only two thousand cases a year of infantile spasm, the company continued marketing their hormone stimulator for a few additional diseases, like rheumatoid arthritis, and they were successful. According to *60 Minutes*, in 2015, "Medicare was spending half a billion a year on Acthar."

In April 2017, Maryland passed a price gouging law. It empowered the attorney general to indict companies if they "shocked the conscience" by dramatically raising the price of an off-patent drug. The following year, the Court of Appeals ruled the law was unconstitutional, and the Supreme Court did not weigh in. Absent a new amendment to the Constitution, Americans who "shock the conscience" have the inalienable right to gouge.

CHAPTER THIRTY

Are Generic Drugs Safe and Effective?

"Reality is a crutch for people who can't cope with drugs."
—LILY TOMLIN

A *New York Times* article told of employees at the large Wockhardt facility in India who were caught "knowingly throwing away vials of insulin that contained metal fragments." Peter Baker, a man who spent six years inspecting generic factories in India, discovered that, in addition, some of the company's employees were fabricating, backdating, and falsifying. Some of the data in twenty-nine of the thirty-eight plants he inspected was allegedly fraudulent or deceptive.

Wockhardt is India's fifth largest pharmaceutical manufacturer, has 8,600 employees worldwide, and is not the only company where shortcuts and violations were detected. In 2014, the FDA sent cautionary letters "to companies operating plants in Australia, Austria, Canada, China, Germany, Japan, Ireland, and Spain." They discussed "manufacturing and packaging violations, testing, quality checks, data collection, and contaminated products."

I'm not sure all brand-name medications are safer than generics. According to Consumer Reports, "Many brand-name drugs are produced overseas, often in the same plants as the generic equivalents."

In her excellent book, *Bottle of Lies*, Katherine Eban tells the story of how one of the large producers of drugs in India entered the American market, then cheated and deceived, and ultimately went under.

The company in question, Ranbaxy, was incorporated in 1961 by Bhai Singh, a member of a wealthy Sikh family. It initially distributed cheap Japanese pharmaceuticals and it had one factory that "reformulated bulk drugs into tablets and capsules." In 1968, the company created and marketed a generic version of Valium, and the drug became a best seller. The sons of the founder, who were studying in the U.S., came home. Over the next twenty years, they were welcomed into the family business. They eventually started fighting over who was in charge.

In 1995, Ranbaxy entered the U.S. market and seemed to be playing by rules set down by the Food and Drug Administration, the FDA. The U.S. agency inspects and checks manufacturers, but they also expect companies to police themselves, keep records, and submit them. The rules for good manufacturing practices were created in the early 1960s. Each critical step in the production of a medication has to be performed and documented competently. Each batch of drugs needs a record. The policy calls for "preventative and corrective action." Ranbaxy seemed to be compliant. Not that the company had the same ethos as their competitor Cipla. A Gandhi-inspired pharmaceutical company, Cipla was clean, well run, and was devoted to supplying medicines to the underserved. Ranbaxy was in business to make a profit. But they were competent and efficient, and the company grew in size and importance. Then they hired an expert to computerize their data. He took his job seriously, and the company started to unravel.

By 1994, half the drugs that were prescribed in the world were generics. Each establishment that supplied drugs to the U.S. was supposed to be checked every two years. The FDA didn't have the resources to fully comply so they employed a model that relied in part on the risk. They used a company's inspection history to identify the facilities they needed to scrutinize. Foreign offices did some of the inspections. Between the years

2000 and 2008, the FDA conducted over a thousand domestic inspections a year and fewer than four hundred checkups in foreign countries.

Companies that produce a drug and sell a medication in the U.S. need to tell the FDA the name and business address of the manufacturer. In 2009, over a thousand generic plants applied to the FDA for permission to sell their drugs here. Forty-three percent of the plants were in China and 39 percent in India.

One of the problems that doomed Ranbaxy was the way the company handled their version of the acne medicine Accutane. Their product was called Storet. Ranbaxy was ready to market their drug when further testing showed the generic formulation wasn't absorbed appropriately. Small quantities of the drug had worked in the lab, but when the company started producing it in quantity, it "was failing." Ranbaxy should have delayed its launch, kept it off the market until they figured out how to correct the problem. But they decided to sell the defective medication to the public while the company's scientists were trying to figure out what they were doing wrong. The company "concealed the problem from regulators." They memorialized their findings and decisions in a dossier that was labeled "DON'T SHOW TO THE FDA." That decision would come back to haunt them.

As Eban explained in her book, drug companies were expected to emphasize quality over cost. "Manufacturing processes had to be transparent, repeatable, and investigate-able." There apparently is always some variation between batches of a medicine, so companies are expected to test each batch and keep real-time records of each drug-making step.

At one point, Ranbaxy was getting big, and it was hard to keep track of what was happening in their factories. That's when they hired a data expert who was born in India. He learned his trade while working for the American drug manufacturer Bristol Myers Squibb. His name was Dinesh Thakur, and he took his job seriously. Before he could computerize figures from the company's many factories, he had to collect the information. He sent his assistants to the facilities, and one by one they came back empty-handed. Ranbaxy, it turned out, had been faking the data that "showed" each batch was tested. They were telling the FDA they were regularly checking the drugs that came off their assembly lines. They were claiming they had proved that each batch was "properly formulated, stable, and effective." But half the information submitted to American and European regulators was bogus. Drugs sold to people in India were not tested for stability and bioequivalence. Adverse events were not reported to the FDA. The drugs may or may not have performed appropriately. The data that said they were up to snuff looked impressive, but it was phony. Drugs made for third-world countries were formulated using relatively impure, cheaper ingredients. Sometimes, instead of testing the products they made, workers at their factories would crush and test brand-name drugs and backdate documents.

In 2004, after his team gathered proof, Thakur reported to his boss, Raj Kumar. It took a little convincing, but eventually Raj was persuaded that the company was systematically faking data. The board of directors had to be notified. Raj created a PowerPoint explanation of what was happening. He believed the drugs that had been misrepresented had to be pulled from the market and the company had to reapply to the FDA.

He made his presentation to the people in charge. It was called a self-assessment report (SAR). When Raj finished, his statement to the board was met with silence. The company leaders didn't seem surprised and weren't willing to admit they were wrong. They would not reapply to the FDA. Kumar was told to destroy his slides and presentation—the SAR. As a result of the leadership's refusal, Kumar resigned in October 2004.

Thakur stayed but was troubled by many of the company's products. He worried about the HIV-fighting drugs that were being shipped to Africa. "He knew they were bad, degraded easily, and would be useless in the heat of Sub Saharan Africa." Two years after he joined the company, Thakur resigned, but he remained troubled. Four months later, Thakur assumed a pseudonym and started writing, initially to the World Health Organization, then to several FDA officials, and ultimately an FDA commissioner. He told them Ranbaxy was faking data—that they were fabricating information to support stability. At the time, whistleblowers in India were sometimes killed. Thakur felt he had to hide his identity. But he also felt compelled to act because he believed "executives at India's biggest pharmaceutical company had committed intentional global fraud." The people at the FDA who received Thakur's emails didn't know what to do. Ranbaxy was a huge company with multiple factories and paperwork that seemed impeccable.

Sometime during the subsequent year, the company was checked out by an FDA examiner who, apparently, didn't know about the allegations. In India, companies were always notified when the FDA was sending an inspector. Visits were scheduled and manufacturers had days to make their facility spotless. The examiner believed the company was honest and honorable and he spent his days identifying and pointing out deficiencies. The factory got a passing grade and the people at Ranbaxy thanked him and promised to fix the problems.

By February 2006, having received multiple surreptitious emails from Thakur, a few people in the FDA had grown suspicious. The Ranbaxy factory was re-inspected. This time, two hard-nosed investigators surveyed the plant. They found deficiencies, and the FDA stalled one of the company's new drug applications.

Eventually, on February 14, 2007, officers carrying guns and wearing bulletproof vests raided Ranbaxy's New Jersey facility. They carried away hard drives and a copy of the company's internal report on their Accutane drug. It was labeled "DO NOT GIVE TO FDA." It proved someone in the company had lied, but it didn't provide evidence of systemic fraud. FDA investigators tried to contact Raj Kumar, the man who gave the PowerPoint presentation, and he evaded them. A company lawyer warned Raj to "be careful about what he tells the FDA because he had exposure." Some leaders at the FDA found it hard to believe that a company with manufacturing plants in eleven countries and sales in 125 nations was systematically providing fraudulent data.

About that time, Tsutomo Une, a leader of Japanese pharmaceutical house Daiichi Sankyo, contacted the head of Ranbaxy and spoke of a joint venture. His company wanted to grow. At the time, Japanese drug manufactures were (and still are) known to shine in the field of quality control. The leaders of the two firms met a number of times. Within four months, a merger of sorts seemed likely, but Une was hesitant. It was public knowledge that the Indian company had been raided and had received warning letters from the FDA. Malvinder, head of Ranbaxy, told Une that the raid was really incited by Pfizer. They were trying to get back at Ranbaxy because the Indian company had "prevailed in the Lipitor patent litigation."

Une believed him, but Malvinder was worried. His lawyers learned that when the FDA raided the New Jersey Ranbaxy headquarters, the 2004 SAR was part of the data they hauled away. It was a smoking gun. It showed that Ranbaxy had been faking the data they presented to the FDA. Malvinder fretted but didn't tell his Japanese counterpart.

In 2008, Daiichi Sankyo became Ranbaxy's principal shareholder. That September, the FDA barred the import of thirty drugs from two Ranbaxy plants. In early 2009, the FDA formally issued an "AIP." The letter notified the company that the agency believed it had evidence that the company's applications had been "fraudulent or unreliable." Malvinder was forced to resign as head of Ranbaxy. He paid a large fine to the FDA, but he avoided criminal prosecution. At a later time, he was sued by the Japanese and was forced to pay them $550 million for overstating the value of his company. Ranbaxy agreed to plead guilty to three felony counts of violating the federal drug safety law and four counts of making false statements to the FDA. The company also admitted that they failed to complete the proper safety and quality control tests on several of the drugs that were manufactured in the Indian factories. We probably would not have learned that Ranbaxy was faking their data had it not been for a troubled whistleblower.

In the years since, the FDA's presence in India has gradually expanded. At the time of this writing, they claim to have more than a dozen full-time staff. Inspections are (officially) frequent and increasingly unannounced. If the agency finds problems, it issues a Form 483, a notice outlining the violations. If violations are not resolved, it can lead to a warning letter and, in the worst case, a ban. Violations range from hygiene, such as rat traps and dirty laboratories, to inadequate controls on systems that store data, leaving it open to tampering.

Eban's book goes on to tell of a few other companies that cheated. In the early 1990s, an FDA inspector observed Sherman Pharmaceuticals of Louisiana burning medication that was returned due to contamination. The company was penalized in 1995 and went out of business.

In 2007, the anticoagulant heparin was manufactured by the American company Baxter, and it made kids who were receiving kidney dialysis sick. The substance that caused the harm turned out to be a contaminant in the Chinese chemicals that were used to make the heparin.

In 2012, Congress passed the FDA Safety and Innovation Act. It directed the FDA to inspect foreign facilities as frequently as domestic plants.

The next year, "the Indian government approved the addition of seven new FDA drug investigators. That brought the total number of Americans checking their facilities to nineteen. That year the number of inspectors in China went from eight to twenty-seven."

In 2014, the FDA conducted more than eight hundred inspections in the U.S. and an equal number of facilities abroad. By 2020, they had inspected half of the 965 foreign production facilities at least once.

Facilities in Europe apparently are rarely inspected by our FDA. Thanks to the 2012 Food and Drug Administration Safety and Innovation Act, the FDA turned over the inspection of meds made in Europe to the inspectors of each country.

In the EU, routine inspections are performed using a risk-based approach and when there is suspicion of noncompliance. National competent authorities are responsible for inspecting manufacturing sites located within their own territories. Sites outside the EU are inspected by the state where the EU importer is located or by the manufacturing country's examiners.

There may be more than one responsible authority for products that are imported directly into more than one member state from a manufacturing site outside the EU.

In 2003, the U.S. accounted for 44 percent of global industry sales, for a total of $216.4 billion. There are a few pharmaceutical companies that participate in both the branded and generic parts of the industry—and the drugs we take have become commodities. Pharmacy benefit managers choose the brands our insurers provide on the basis of price, quality, reliability, and kickbacks. No one knows exactly how they make their decisions.

Based on the Ranbaxy story and numerous interviews, Eban asks a very important question: Should brand-name and/or generic pharmaceuticals be trusted? It's an especially important question because more than

40 percent of finished drugs used in the U.S., and 80 percent of active pharmaceutical ingredients, are produced overseas. Pharmaceutical factories are scattered around the globe, there's a lot of money involved, and people sometimes cheat.

Deception by major players, sadly, is not confined to the generic pharmaceutical industry.

- In 2015, the environmental protection agency discovered that 11 million diesel autos made by Volkswagen contained a special device that allowed the company to intentionally lie about their emissions. The company admitted guilt and paid a $25 billion fine.

- In 2008, the former head of NASDAQ, Bernie Madoff, admitted he had "conned his investors out of $65 billion over many decades." He went to jail.

- More recently, airplane manufacturer Boeing fitted some of its newer planes with large engines and special software but failed to train some of the pilots. Their actions led to the crash of two aircraft and the loss of hundreds of lives.

- Early in this century, huge trusted banks sold collections of mortgages whose worth was falsely inflated. When housing prices started dropping in 2007, the values of these mortgages plummeted and the banks didn't have enough cash to avoid a run. The economy was saved by a $16.8 trillion government bailout.

When I was a practicing doc, pharmaceutical salespeople came around with donuts and both promoted their brand-name drugs and questioned the safety and reliability of generics, so doctors of my era were periodically brainwashed.

Are generic drugs safe and effective? My take: The world has bad actors and cheaters, but most companies and inspectors take pride in their work. They don't want to let their friends and neighbors down, and they strive to avoid creating or choosing an inferior, worthless, or harmful product.

CHAPTER THIRTY-ONE

The FDA—The Fox That's Guarding the Henhouse

"The protection of a man's person is more sacred than the protection of his property."
—THOMAS PAINE

The FDA protects us from harm and misrepresentation and it makes sure that newly approved drugs are reasonably safe and at least temporarily effective.

If some critics were correct, if it once took too long for new drugs to be evaluated and approved, that's no longer the case. If anything, the FDA is tilted in the other direction.

Drugs are often approved when their long-term effect won't be known for years—like the medications that lower the A1C, the marker of good blood sugar control in a person with diabetes. Tight control may or may not lead to a better outcome. When the average blood sugar is close to normal, there's less damage to the small blood vessels of the eyes and kidney. But in a study of the frail elderly, an emphasis on keeping the average blood sugar low led to "increased mortality and did not significantly reduce major cardiovascular events." Hypoglycemia, a blood sugar that's very low, can cause irrational behavior, falls, and even death.

Cancer drugs are marketed when they cause a cancer to shrink and give people an extra few months of life.

Physicians don't have to wait for FDA approval before they prescribe an experimental medication. There's a mechanism that allows physicians to legally use a drug that is still being tested. Called the IND (Investigational New Drug) permit, the process has been around since 1987. If a therapy is needed urgently and a manufacturer has an experimental product that might help, a doctor can apply. The FDA received more than five hundred commercial, research, and emergency IND requests in 2018 and in 2019.

Using a different tool, the Food and Drug Administration allowed physicians at Walter Reed to treat President Trump with the antiviral drug remdesivir available under an "emergency-use authorization."

In 1992, after Congress passed the Prescription Drug User Fee Act (PDUFA), two-thirds of drug approval expenses were paid by big pharma. The fox was paying the regulators who were guarding the henhouse.

The FDA uses a number of advisory boards, groups of physicians who are experts in the field. The FDA officer makes the final decision, but, in tough situations, it must be nice and at the same time awful to have a group of M.D.s who serve as a sounding board and buffer.

In 2007, they used the advice of boards of doctors when they evaluated Avandia, a "glitazone," and the doctors blew it. The drug they evaluated was made by SmithKline Beecham and it did lower the blood sugar, but it also raised the level of cholesterol in the blood. The other glitazone the board evaluated, Actos (pioglitazone), didn't worsen blood lipids. In 1999, the agency approved both rosiglitazone (Avandia) and pioglitazone (Actos) for use in people with diabetes. Doctors could prescribe either, but the FDA wanted companies to monitor the drugs for problems.

By 2006, both drugs were grossing more than $1.5 billion a year and it was clear that Actos did not increase the risk of coronary disease, but Avandia did. It also sometimes caused heart failure and fluid in the lungs and legs.

FDA panels were convened and the "experts" voted to keep Avandia on the market, but they added black box warnings to the packaging. Then a 2007 medical study convincingly showed that Avandia caused heart disease. Drug sales dropped, but a million prescriptions a year were still being written.

The panels of experts convened by the FDA agreed that Avandia "posed significant cardiovascular risk," and twelve of the thirty-three panel members thought the drug should be removed from the market.

The chairman and nine others voted for much stricter controls. The doctor in charge wrote that "several meta-analyses revealed a significant increase in the risk of myocardial ischemic events among patients taking rosiglitazone. . . . But a second analysis failed to demonstrate a similar risk." Then he added a little gibberish: "The results regarding the safety of rosiglitazone raised new questions about relative and absolute risks."

In July 2010, the manufacturer of Avandia settled a lawsuit for $460 million for the harm the medication did. Compared to revenue of $1.1 billion the prior year and much more in the years before the 2007 hearings, the monies paid did relatively little harm to the company's bottom line. Pioglitazone—Actos—is still being widely used. According to the website UpToDate, rosiglitazone is still being marketed, but it is "rarely used and no longer widely available."

At the height of the AIDS epidemic, activists protested the delay between a new drug's submission and approval. In response, the agency created fast-track rules that sped up the development, assessment, and sales of new treatments for life-threatening conditions. The FDA also made unapproved drugs available to people who had AIDS (and other serious conditions) who were unable to enroll in clinical trials.

In the 1980s, the agency created "treatment INDs." To obtain a medication, a provider has to submit a form that can usually be filled out in forty-five minutes. The agency then takes up to four days to process a non-emergency application. Emergency requests are approved in less than a day. Between 2005 and 2014, twelve hundred forms were submitted each year and over 99 percent were approved. Most were for a single patient, and half were for an emergency. In 1970, a strong demand for experimental cancer drugs led the FDA to adapt an early-access policy. In 1992, the agency started allowing speedy approval on the basis of end points that were seen as "reasonably likely to predict patient benefit."

During the 2014 African Ebola outbreak, acting on preliminary data, the FDA authorized the use of six tests that rapidly identified infected patients, and they reviewed IND applications for two investigational vaccines in less than a week. Then they allowed developers to proceed with phase one clinical trials.

In 1992, Congress passed the Prescription Drug User Fee Act. It authorized the FDA to collect money from pharmaceutical manufacturers and told the FDA to review special drug applications within six months. Ordinary applications had to be assessed within a year.

As therapies were developed and authorized more quickly, the FDA started "requesting" post-approval studies. Under the 2007 FDA Amendments Act, Congress allowed the FDA to "require" studies after a drug was approved.

In spite of the law, only half of the post-approval studies were completed within five years: "20 percent had not been started; and 25 percent were delayed or ongoing."

CHAPTER THIRTY-TWO

The For-Profit Insurance Companies Take Over

"It's difficult to get a man to understand something when his salary depends on his not understanding it."
—UPTON SINCLAIR

I n 1940, after conquering most of Western Europe, Hitler mercilessly bombed British cities and tried to bring England to its knees. Eighteen thousand tons of explosives were dropped in eight months, and close to forty thousand people were killed. The British government began paying for ambulances and for the care of civilians and combatants who suffered fractures, burns, and head injuries. The ill and injured were transferred to private suburban medical facilities, "doctors received government salaries," and the Ministry of Health "built or expanded hundreds of hospitals." Laboratories and X-ray complexes were updated and three and a half million people were moved to the countryside. Unexpectedly, amidst the death and destruction, most Brits got healthier. When the war ended, the people didn't want government care to end.

In 1948, Parliament formally passed the National Health Service Act and medical care became a taxpayer-funded right. "It started providing and paying for a majority of the British health care needs on the basis of need, not the ability to pay." When Margaret Thatcher became prime minister in the 1980s, she opposed welfare and public housing, but "the National Health Service remained the critical mainstay service for the British public."

Postwar France was devastated. "Many people were too destitute to afford heat, let alone medications or hospital visits." When they got private medical care, the French usually paid cash. However, a little over a decade before the war started, the government had established health insurance for "salaried workers in industry and commerce whose wages were low enough." It was like our Medicaid. In 1945, government insurance was extended to all industrial workers and their families. Throughout the 1960s, more and more groups were added. Finally, in 1999, the country instituted universal health coverage.

The French system is funded by social security and, best I can tell, covers most of the care needed by people of all ages. Each year, the French parliament determines the maximum amount the government will spend for health care in the coming year. In addition, there are a number of voluntary funds and they pay for 15 percent of the medical care people receive. Most French are apparently covered by one.

In France, public facilities provide the majority of the hospital beds, and they transfer people with complex illnesses to one of thirty-two regional facilities. Fifteen percent of the inpatient care is supplied by hospitals that are owned by religious organizations, foundations, or mutual insurance associations and 10 percent of the hospital beds are located in private for-profit institutions.

In the U.S., President Franklin Roosevelt didn't mention health care or call it a right in his famous 1941 speech to Congress. The four freedoms he championed were speech and expression, "the freedom to worship God in a person's own way, freedom from want and freedom from fear." He had wanted federally sponsored health insurance to be part of the 1935 Social Security Act, but "he allowed it to be thrown out in order to hurry the bill through Congress."

During the Second World War, the military and VA cared for the wounded. Wages were restricted as part of the 1942 Emergency Stabilization Act, and employers attracted workers by offering health insurance. Our nation's work-related system created inequities, and it helped foster the rugged individualistic belief that people don't deserve health care. They have to earn it.

In 1947, medical costs were rising and health care was too expensive for many. That January in his State of the Union address, President Truman said: "Of all our basic resources none is more valuable than the health of our people." He proposed a plan that provided medical care to all who needed it. "A government funded health care program that everyone pays into and they can withdraw from when they are ill. Not a charity, but (care) on the basis of payments made by the beneficiaries of the program."

Truman's idea initially had widespread support. Then the powerful white male American Medical Association (AMA) launched a massive campaign: Radio and newspaper ads, pamphlets and mailers told voters to "keep politics out of medicine." "Truman was a Communist." (At the time, Communism was a real scare.) "His plan was socialized medicine. Government officials will intervene in medical decisions and destroy the sacred doctor patient relationship."

Popular support plummeted and the bill failed to get through Congress. Efforts by subsequent politicians to provide government-sponsored health insurance were repeatedly opposed by the "influential" American Medical Association.

In 1951, Aetna and Cigna began marketing medical insurance that tended to target people who were in good shape and unlikely to need expensive care. The policies advanced the belief that the cost of medical care should not be a responsibility that is shared by young and old, healthy and infirm. Many young and healthy people began to believe that inexpensive quality health insurance was one of their rights.

In 1960, John Kennedy, the son of a wealthy Irish immigrant, became the nation's youngest ever president. The year before he was elected, Cuban revolutionaries ousted their ruler, a dictator named Batista. Fidel Castro became the island country's leader. Boatloads of Cuban merchants and land owners fled and became residents of Florida.

Castro was a Communist. He "broke up large landholdings and gave them to the peasants. The Cuban state took possession of foreign oil company holdings, and foreign-owned property."

A few months before Kennedy was elected, Castro traveled to New York City to attend a session of the United Nations. He stayed in a hotel in Harlem. While he was there, he invited the leader of Russia, Nikita Khrushchev, to come over. At the time, the U.S. and Russia were adversaries.

In response to Castro's invitation, Khrushchev's bodyguards garnered a police escort and Khrushchev climbed into a car. The two leaders met in front of the hotel. Castro, a tall man with a beard and "a face that was both pleasant and tough, bent down." The two hugged in front of a gaggle of reporters, and Russia became one of the Cuban government's major supporters.

The thought of a Soviet ally a mere 110 miles south of Florida troubled some U.S. leaders. In 1961, a group of former Cubans sponsored by the CIA invaded the island. The U.S. military didn't get involved, and the intruders were quickly defeated and captured. After the fiasco, the popularity of the U.S. president waned.

Early in his presidency, Kennedy tried to change the system that kept southern blacks from voting, and his efforts went nowhere. He was also unable to convince Congress to pass legislation that provided medical care to our older citizens.

Then, in the fall of 1963, Kennedy was assassinated. The nation mourned, and Lyndon Johnson, former leader of the Senate, became the nation's president. Under his direction, Congress paid tribute to their slain leader and passed a law that established Medicare and Medicaid. The entities were funded by a payroll tax, an additional 2.9 percent of a person's earnings that were taken from each paycheck. (It was never called socialized medicine.)

In 1964, a year before it enacted Medicare, Congress passed the Civil Rights Act. One of its provisions, Title VI, prohibited discrimination on the basis of race, color, or national origin in programs that received federal funding.

The following year, when Medicare was being debated, there was no mention of Title VI. It didn't come up and no one wanted to raise it. "The assumption was that accommodations would be made and they would allow segregation to continue on a separate but equal basis." The nation's response to the 1954 *Brown v. Board of Education* Supreme Court decision had shown that when it comes to race, changes don't happen very much or very fast. The 1954 ruling had ordered states to put a stop to segregation in schools with "all deliberate speed." But "the vagueness about how to enforce the ruling gave segregationists an opportunity to organize resistance."

In 1956, two years after the Supreme Court decision, the University of Alabama enrolled the school's first black student, Autherine Lucy. When "riots engulfed the campus," she was expelled for "her own safety." In 1962, demonstrations greeted James Meredith, the first black student to attend the University of Mississippi. He was a veteran of nine years in the U.S. Air Force and a man who felt it was important to make a statement. He was ready. I started medical school at Washington U. in St. Louis in 1958, four years after the *Brown* Supreme Court decision. During my four years as a student, the school's main hospital (Barnes) provided care for "a number of African-American patients who had interesting cases or could pay for treatment. They were bedded in inferior wards located in the basement amid the pipes." The school didn't graduate its first black physician until 1962. Berkeley, California, the first sizable city with a "substantial portion of black students," waited twelve years before it started voluntarily transporting some white children to schools in black neighborhoods and some black students to schools that were largely white.

"The Medicare law might carry the threat that federal funding could be withheld from any hospital that practiced racial discrimination." That's what Title VI of the Civil Rights Act "required." But the federal office that was in charge of federal funding based on the presence or absence of discrimination only had five employees. Swift action seemed unlikely. "Hospitals might have to come up with a plan, but they could presumably proceed gradually and cautiously. Ultimately racial changes wouldn't get done and hospitals would operate on a business as usual basis."

After Medicare became law, the HEW (Health Education and Welfare) secretary, John Gardner, put out a quiet call for volunteers. More than one thousand people from federal agencies and tens of thousands of civil rights organizers offered to help. As David Barton Smith explained in his book "The Power to Heal," "They wouldn't let anybody off the hook. The reason they were so successful is that they had this secret army of local civil rights workers and local health workers making sure [the hospitals] complied. Hospitals quickly figured out that they couldn't fake it, "and racial discrimination in hospitals soon ended."

The 1965 law gave a new set of rights to citizens over sixty-five and the very poor. Medicare paid for the hospital care and some home services for citizens over sixty-five, and Medicaid covered the care of citizens who were "poor enough"—whose "resources were insufficient to pay for health care." Medicaid also covered nursing home fees for people who met the eligibility requirements. Fifty to 75 percent of the program's costs were paid by the feds, and each state paid the rest. States with higher per capita incomes paid a larger percent. For people who earned up to 100 percent of the Federal Poverty Level (FPL), the law covered the care of pregnant women and infants (up to one year of age) as a state option.

No one pointed out that the programs were socialized medicine and that most people liked them.

In the years that followed World War II, Blue Cross and Blue Shield were started in one state, then another. They were public, tax-exempt corporations. By 1945, the "Blues" were responsible for two-thirds of the nation's hospital insurance, and by 1955, 60 percent of Americans had an insurance policy.

During the subsequent decade, medical care became more effective, technical, and hospital based. Costs escalated. In the early 1970s, the political elite felt there was a "crisis," and in 1973, during the Nixon presidency, the Health Maintenance Organization (HMO) Act was passed. The idea was to get groups of doctors working together. In addition to treating illnesses, physicians were now being asked to keep people healthy. Companies with more than twenty-five employees were required to offer this new entity as an option.

Groups popped up all over the country, and they took customers away from the "indemnity" insurance companies. In response, traditional insurers started buying HMOs. The new bosses put pressure on doctors and hospitals. They tried to modify the way medicine was practiced by enforcing "guidelines." These are recommendations that were developed by authoritative medical committees. They help doctors decide what tests to perform in certain situations. Physicians were paid less. Knowing they would only be compensated for a portion of their charges, doctors and hospitals raised the "list price" of their surgeries, visits, and procedures.

At one time, HMOs tried to kick women out of the hospital the day after they delivered a child or had a breast surgically removed. Some of the patients were ready to go home, but others had medical or physical problems that created risks. An outcry led to publicity and lobbying. Laws were passed, and the practice ended.

The law restricted the amount a sick person in an HMO had to pay in the form of co-pays and deductibles, and HMOs were required to provide insurance "without regard to health status." The organizations were required to have an annual open enrollment period, during which people with preexisting conditions could become members.

For-profit companies could exclude high-risk individuals. A law passed in 1945, the McCarran-Ferguson Act, had "exempted the business of insurance from most federal regulation." For-profit companies were able to cherry-pick—to offer the young, healthy, and employed low-cost policies. They were able to reject individuals who had medical problems, and they did. It was hard to turn a profit from policies that provided care to people with expensive needs.

HMOs could not offer companies high deductible or low benefit policies, but for-profit insurers could and did. Employers increasingly bought the limited policies from the for-profit insurance companies.

By the 1970s, socialized medicine and the government and taxes paid for the health care of a large chunk of the populace. Medicare and Medicaid paid for the insurance policies of the elderly, individuals with chronic renal disease, and people who earned very little money. The feds funded the Veterans Administration, the Indian Health Service, and prisons.

There are 8 million federal employees. More than 70 percent of their health care premiums are paid for by the government.

More and more of the higher-risk individuals could only get health insurance from the "Blues" and from HMOs like Kaiser of Northern California. The nonprofits had increasingly fewer low-risk individuals to help balance the cost of care.

But Blue Cross and Blue Shield still controlled a large part of the health care market. The Blues weren't a single large company. They were a conglomeration of state-by-state plans that were nonprofit and tax exempt. Their boards and CEOs were very well (and sometimes quite richly) compensated, but they had no shareholders. They didn't need to earn money or pay dividends, and the money they took in was actually used to pay for medical care. Driven by "need," not "greed," these companies only charged what they required to cover their costs. By so doing, they established a de facto price boundary. It limited the amount the for-profits could charge for policies.

In 1993, Bill Clinton became president. The premiums people paid were going up 8.5 percent a year. Many Americans were dropping their coverage and polls indicated the public would welcome controls.

Clinton decided to tackle the "health care crisis." "He asked his wife Hillary to create and lead a task force and take some of the heat and she accepted."

As Hillary later put it, "I wanted to try to make a difference in people's lives. What could be more important than solving our nation's health care crisis?" She interviewed people who had been denied coverage and she publicly unloaded on the industry: "It's almost like they decided people should just drop dead without ever going into the hospital."

Her goal, she told Congress, was "guaranteed universal health coverage at an affordable cost for every American. She proposed a comprehensive package of benefits and health care that could never be taken away."

In 1990, while Hillary Clinton's group was working on their bill, the CEOs of insurance companies pow-wowed with drug and medical device manufacturers and with the people in charge of hospitals. They formed a group and called it the HLC, the Healthcare Leadership Council.

I suspect the Clintons thought their proposals would be popular and were not prepared for the attack and spin that followed.

The plan Hillary and her medical advisors came up with sounds a lot like Bernie Sanders's Medicare for all. It would have required health plans to provide an all-inclusive benefit package that would have been regulated by a regional nonprofit organization or an agency of the state. When people got their insurance from their jobs, the employer would need to pay 80 percent of the cost of the premiums.

Hillary's proposal was comprehensive. Hospital, outpatient, drug, and dental care were part of the scheme. To keep costs from going through the roof, there was a regional money target. If it was exceeded, the payments to hospitals and doctors would drop. Out-of-network emergency charges couldn't be inflated. Balanced billing—making the patient kick in when the insurance company refused—was prohibited. There were caps on premiums. Drug prices would have been "reviewed," and a tax increase was not recommended.

Bill Frist, physician, senator, Republican, and founder of a huge for-profit hospital chain, said, "Hillary Clinton was seen as more rigid, probably more principled." She was out front on the issues that "maybe half the country didn't think it should go that far," so she was easy to demonize. One of the opposition ads talked about a whole bunch of guys who were in Hillary's testicle lockbox. Bill was portrayed as Hillary's poodle. Rush Limbaugh pulled out a picture of her as an evil witch and showed photos of her daughter.

The HLC claimed that if the proposals became law, health care would be rationed. Patients' rights would be compromised. The Clinton plan "was formed by academic elites behind closed doors."

TV ads featured Harry and Louise, two fictional Americans who thought the suggestions would let "Washington bureaucrats decide how much care you and your family receive."

Congressmen and -women were lobbied, there were letter-writing campaigns, and the Medicare program was pronounced "dysfunctional."

Insurance companies argued that the "free market could work if government got out of the way."

The effort to prevent health care reform cost $300 million, and in the end, the Clinton proposals were not accepted by Congress. Hillary Clinton went on to fight other battles.

"The Blues were hemorrhaging money," and the HMO I worked for was in grave risk of failing. Then one day in 1994, I heard a group at the nurses' station discussing an article in the local paper. The board of directors of Blue Cross had resigned. They formed a for-profit health care organization and took all the low-risk policies with them.

We now know that in 1994 the national Blue Cross and Blue Shield board of directors met in Washington, D.C., and voted to allow its members to "operate as for-profit companies." Blue Cross of California changed its name to WellPoint. They became a new, for-profit company and took a bulk of the policies that insured low-risk individuals. They then bought the nonprofit Blues plans in Missouri and Wisconsin. Blue Cross of Indiana became Anthem, and merged with WellPoint. By 1996, California, Colorado, Ohio, Virginia, Kansas, Maine, Missouri, and Georgia had changed over. There were a few remaining not-for-profit insurance plans, but the private companies were in charge and could now call the shots.

The for-profits were not required to "provide affordable, accessible health care" to all the people of their state. They could, instead, provide and/or charge for health care according to risk. They could concentrate on making money for their stockholders. Kaiser, the group I worked for, modified its business model.

In 1987, the for-profit companies contended they were providing a service to the community and convinced the IRS to give them "special tax benefits."

In 2008, a former insurance executive named Wendell Potter had enough, quit, and wrote an "exposé," a book titled *Deadly Spin*. His revelations didn't make headlines or shake events much, but his story is similar in some ways to that of the man who exposed big tobacco. Formerly one of the nation's better PR people, he was a "spin" expert. When the press came after his company for misdeeds, questions were not answered. He taught his people to respond with talking points. CEOs were trained and Potter attended interviews and limited the scope and duration of the sessions.

In his book, he explained the "right way" to reply to questions about the 45 million Americans who didn't have health insurance. Spokesmen and -women were taught to blame the victim: 40 percent of the uninsured were young adults, and company representatives explained that many of the uninsured believed the risk of injury or illness was too low to justify the cost of the premium. Thirty-five percent earned $50,000 a year and should be able to afford insurance, so it was a question of choice. Twenty percent were not citizens.

Potter, a poor boy from the back country of Tennessee, was the first in his family to go to college. His father served in Europe and North Africa during the Second World War, then came home, married, and opened a grocery store. When the business lost money, his dad got a job in a "brutally hot factory," and toiled there until he retired.

In and after college, Potter was briefly a journalist. Later, he became a press secretary for a man who ran for governor and lost, and he next became a "spokesman for a failing banking empire." In time, the health insurer Cigna hired him "to help boost awareness" of the company's health care business and he became a PR man extraordinaire. In addition to helping devise the messages of some of the nation's leading insurance companies, he participated in the demolition of Bill Clinton's health bill. Well paid, important, and at the top of his game, he paid a visit to his parents in July 2007. While there, he learned that John Edwards, the presidential candidate, was scheduled to talk about health care reform at a fair in nearby Wise County, Virginia. Wendell was devising talking points for his company and looking for inspiration, so he decided to attend and listen.

The health fair was produced by an adventurer and TV personality named Stan Brock. In 1985, Brock founded a nonprofit organization called Remote Area Medical. "Volunteer doctors, nurses, technicians, and veterinarians, at their own expense, were taken on expeditions where they treated hundreds of patients a day under some of the worst conditions." Crews worked in Katrina after the hurricane, Haiti after the earthquake, and held daylong clinics in underserved areas in the U.S. Wise County, Virginia, was one of those locations.

Potter arrived early and parked in a jammed lot. Some of the people who formed long lines had slept in their vehicles. It had rained the night before and it was damp. Inside the treatment area, large numbers of people waited in lines that led to clinics. Dentists pulled teeth. Nurses performed Pap smears. Potter recalled mammograms, sigmoidoscopies, and doctors cutting out skin tumors. Some of the people got their care in tents or animal stalls.

Potter was troubled by what he saw. This wasn't happening in a third-world country. This was America, the land with the best health care in the world. Two-thirds of the people at the health fair had no insurance, but a third did. Problem was, many of their policies didn't kick in until thousands of medical dollars were paid by the policy holder. In some cases, we're talking about $15,000 each year. In other cases, the amount was $30,000. The average income in the nearby counties was $23,000 to $26,000. A "family health insurance" policy went for $13,375. (Sixty-two percent of bankruptcy filings in 2007 were the result of health care costs.)

What Potter witnessed could have been spun. The young uninsured were risk takers. People who earned $50,000 a year were shirking their responsibilities. Noncitizens would be lumped and denounced as illegal aliens, though many were students. Other noncitizens were in the country legally, and were performing jobs Americans couldn't or wouldn't do. They didn't have health care because they couldn't afford it.

Potter knew how to tell the story, but when he looked into the faces of the people standing in line in the rain for hours to get care in animal stalls, he saw folks who could have been his relatives or neighbors. He didn't know them, but they seemed so familiar. They were like his dad and mother. The army had sent Potter's father to Europe and North Africa during World War II. When his dad returned, both parents dropped out of high school and worked. "They sacrificed years to send Wendell to college." The people Potter was being paid to vilify were no longer anonymous or faceless.

A few days later, Wendell was aboard a company jet. When his food arrived on a gold-trimmed plate, he thought about the people who, days earlier, were receiving medical care in animal stalls. That's when he decided to quit the spin game. The contrast between the compensation executives exacted and the care they denied was something that Potter pointed out in his book.

When Edward Hanaway left Cigna in 2009, he wrote a book blaming the health care consumer for the rising cost of health care. His retirement package was valued at $111 million. In 2007, a top average health insurance CEO had a salary of over $11 million. William McGuire of UnitedHealth backdated his stock options, was caught, and paid back $620 million. He then managed to survive on $530 million in nonstock compensation plus an additional $800 million in stock options.

In his book, Potter documented, but avoided, direct comments about the pay executives received, dollars that could have gone into providing better or more care (or even used as dividends for investors).

Corporations in this country are strange inventions. They are endowed with the same freedoms of speech the Constitution gives to humans. As quasi-people, they can be greedy or they can draw a line when they have enough. Some only seem to have the conscience and ethics needed to mollify the public and the government. They believe their purpose for being, their job, is to make money.

CHAPTER THIRTY-THREE

Obamacare—
The Affordable Care Act

"The people who are crazy enough to think they can change the world are the ones who do."
—STEVE JOBS

B y 2008, a number of insurance companies had instituted policies that troubled Congress and became the topic of exposés and newspaper articles.

It now seemed to matter that people with serious preexisting medical conditions could not buy health insurance.

When insurance companies couldn't raise premiums, they enhanced their profitability by creating large copays. When someone in a small business developed an expensive problem, an insurer would substantially raise the cost of their premiums. They knew that if they charged enough, the client would not renew the policy.

Aetna spent millions for technology that identified establishments that cost insurers more than the companies paid in, and in the early 1990s, they shed 8 million enrollees. Health Net gave bonuses to employees who discovered policies that lost money.

Offering insurance when people are well and taking it away when people need it may seem slimy to some, but to others it's just "business." Health insurance policies were thought of as commodities and they were supposed to be profitable.

Most people, however, were disturbed by a tactic the companies called "rescission." When someone got sick, their original questionnaire was scanned for minor, mostly irrelevant misstatements. Wendell Potter tells of a man who had an angioplasty for coronary disease. The insurer refused to pay because "he failed to disclose" a history of heartburn. His omission gave the company the foothold they then used to avoid paying $130,000 in medical costs. The case was not unique. When it was detailed in a Los Angeles newspaper, letters poured in.

WellPoint, according to Reuters, "singled out women with breast cancer for aggressive investigation." The company found insignificant misstatements in entry questionnaires and earned over a hundred million dollars by canceling the policies of females who had developed breast cancer.

When questioned by a congressional committee, insurance company CEOs were asked if they would end the policy of rescission and they said no, they wouldn't.

In 2006, Massachusetts passed a health care law that became a partial blueprint for Obamacare. The state extended "affordable," compulsory, state-subsidized health insurance to all its citizens at a time when a Republican, Mitt Romney, was governor and it seemed to be working. Vermont followed suit. Some (perhaps Obama) wondered if this could be a model for other states. Problem was, the Massachusetts situation was unique for a number of reasons. When Michael Dukakis was governor (in 1984), the legislature decided to make sure hospitals got paid. The state established the "Uncompensated Care Pool." It was funded by hospitals and insurance companies according to a formula, and it accumulated a pile of usable money. The funds helped protect the finances of people who were recovering from a catastrophic event and there was money left over. By 2003, the year Mitt Romney became governor, the (now called)

Health Safety Net was bringing in $157 million a year and it provided funds the state could work with. Mitt also got help from Ted Kennedy. Through some legislative maneuver, Massachusetts "secured three years of additional Medicaid funding, $1.05 billion."

Massachusetts happened to be one of eleven states that still had a functioning 2.8 million members' strong nonprofit Blue Cross and Blue Shield. The company used 90 percent of the money taken in to care for people, and didn't have any shareholders it needed to satisfy. Medicare had been in existence since 1965. The bureau that ran the program, the CMS, had six thousand employees and had established payment schedules for "more than ten thousand physician services." Prices were adjusted "to reflect the variation in practice costs from area to area." The independent agency that assessed the "worth" of a visit or intervention helped keep a cap on costs.

The Massachusetts program was less comprehensive than Obamacare would be. The Affordable Care Act would pay a large part of the insurance premium for people who earned less than 400 percent of the federal poverty level, and Massachusetts made the cut at 300 percent.

By the time Obama became president in 2009, the ability of medicine to enhance the quality and quantity of most lives had grown and costs had risen dramatically. (According to Obama's 2020 *New Yorker* article, more than 43 million Americans were uninsured, premiums for family coverage had risen 97 percent since 2000, and costs were climbing.)

Over 17 percent of the national GDP was spent on "health," and a majority of Americans had government- or business-sponsored quality affordable health care. At the same time, many citizens were excluded, and the unregulated private insurance companies had grown powerful and arrogant.

I voted for Barack Obama and assumed he would fix health care. It was part of his platform. At the time, he favored the "public option"—like Medicare or Medicaid for those who want it.

The 2008 election gave Democrats a majority in the House and Senate, but Obama believed in inclusion. He tried to work across the aisle and talked about compromise.

The Republicans weren't having it. For some, it was because the nation rejected John McCain, a popular white candidate. Others were probably just sore losers. A few perceived a loss of status because an African American was in the White House. Many assumed Obama viewed the world through the eyes of a black man struggling for a better life for "his people." And he did. During the three years before he entered law school, he "struggled to find his identity." He lived and worked in a black Chicago neighborhood and tried to organize and help the residents. At the same time, Barack was the son of a white mother and he was in his thirties when she died. His grandfather fought in the Second World War, and Obama had no ancestors who were former slaves. His father was born and raised in British-occupied Kenya. The great-great-great grandmother of the woman Barack married was a slave. Michelle, Barack's wife, grew up in a working-class Chicago neighborhood where the schools were good and the white inhabitants started leaving when the blacks moved in.

The 2008 presidential election was the most racially and ethnically diverse in U.S. history. That troubled Beltway Republican politicians. On the evening of Obama's inauguration, opposition leaders met and decided they would be against everything that he proposed.

At the time, the economy was on the edge of another great depression and Barack favored a Wall Street bailout. Democrats voted with him and Republicans didn't. He used the money to prop up the big banks, and decided to NOT punish their leaders. The heads of the banks then gave themselves millions in bonuses at a time when they were taking homes from people with mortgages. Some of the mortgages were owned by people who had lost their job as a result of the recession, and they couldn't make the payments. To many, it seemed like welfare for the privileged and bleak capitalism for the common man.

Obama's popularity declined. It was still his first year in office and he decided to pass a health care bill based on his campaign promises. He sought bipartisan support and Republican senators decided to oppose him. They forced their fellow legislators to just say NO to any legislation Obama proposed. According to *Frontline*, some thought the bill would be the president's Waterloo, and that it would allow them to recapture the Senate.

The legislative process took more than a year. The Senate would not pass a law unless sixty of the one hundred senators voted for it, and there were exactly sixty Democratic senators. Obama needed every vote. The insurance companies didn't want a public option and they wanted to require everyone to buy insurance. The drug industry didn't want Medicare to be able to negotiate drug prices. There were ups and downs, protests, and an expensive insurance company advertising campaign. Ultimately, to get the needed votes, Obama had to "kill" the public option and water down a new, special tax on medical devices and equipment. To obtain the vote of Nebraska's democratic senator, Ben Nelson, the federal government had to agree to pay the full price of his state's Medicaid insurance.

Then Democratic senator Ted Kennedy died. He was replaced by a Republican, and the Democrats no longer had a sixty-vote "super majority."

In the end, Obama campaigned hard, "expended a lot of political capital," and Congress passed the current Affordable Care Act.

In March 2010, Obamacare became law. The Affordable Care Act got rid of rescission. The law prevents insurance companies from denying coverage or charging a higher price to someone with a preexisting health problem. Health plans can no longer set a lifetime limit on how much an insurer has to pay to cover someone, and insurers are required to offer a minimum package of benefits. Preventative health services must be provided without a copayment. Children are allowed to stay on their parents' policies until age twenty-six.

But—in an effort to get enough money into the system, to make it workable, and at the insistence of the insurance companies—the law required all people to buy a policy. It also enacted a tax penalty for large employers that failed to offer affordable coverage, or individuals who failed to obtain insurance.

(In his 2020 *New Yorker* article, Obama said that it was he and his staff that decided that each person should be required to purchase health insurance, the "individual mandate." He wrote that Mitt Romney called the individual mandate "the ultimate conservative idea," because it promoted personal responsibility.)

The law created a situation where low-risk healthy young people were placed in an insurance pool that contained people who were older and sicker. The costs of the young person's premium went up. That didn't seem fair to young people who were raised in a system where the cost of care wasn't shared. People who had to pay more for their health insurance were unhappy.

If a state agreed, people whose incomes were a little above the poverty line were now eligible for Medicaid. Thirty-six states and the District of Columbia signed up shortly after the law was passed, and by 2021 three additional states had joined them. As of January 2021, twelve states were still declining the freebie. Without the extra patients, some rural hospitals failed and people went without care.

The ACA, Affordable Care Act, supplemented insurance premiums for people whose earnings were quite low but who didn't qualify for Medicaid, and it allowed premiums to be sensitive to the marketplace. If a state only had one insurer, it could charge more, and in the face of stiff competition, companies could offer cheaper rates.

Group policy rates were still the result of bargaining, and the size and average age of the people to be insured mattered. High deductibles were permitted. An older person could be charged up to three times as much as a twenty-six-year-old. Tobacco users could be forced to pay twice as much as nonusers.

Prices could not be higher for a new enrollee who had metastatic cancer and was receiving expensive chemotherapy. Nor could more be charged to the person who had hemophilia (and bled easily) or someone who just had a heart attack or a stroke. A person's "current health or medical history"—their preexisting condition—could not affect the premium's price.

The law's approach ran counter to the basic premise of insurance. Companies evaluate risk. They charge more for flood insurance in a zone that is periodically inundated, more for fire insurance for homes in a highly wooded area, and more for earthquake insurance in California. On a societal level, it's not wrong to discourage people whose homes were washed out a few times to stop rebuilding in low-lying areas that flood every few years. But most developed countries don't treat health insurance as a commodity, a product that is supposed to earn a significant amount of money. They don't consider risk-assessing before they decide if they will provide health care, and they don't invoke an illness penalty. They believe that we are all one step away from a serious medical illness or injury. Their young and healthy share the cost of everyone's health care, and they assume they will be cared for when and if they become sick or are injured.

As Atul Gawande put it, a century ago the average American didn't grow old. When someone suffered a catastrophic event like pneumonia, a heart attack, or a bad accident, many died or were disabled but some walked away unharmed. Modern medicine often rescues people who would have died. We cure pneumonia, pin broken hips, and stent narrowed coronary arteries. Subsequently, people may or may not be as healthy as they were, but they now have a "preexisting condition." Before the Affordable Care Act was enacted, these people found it difficult or impossible to acquire insurance.

Insurers call the funds they spend caring for the ill a "loss." If the company uses 20 percent of the money they take in for executives, payroll, and stockholders, their medical loss ratio is 80 percent. The ACA, Affordable Care Act, capped the amount companies could keep at 20 percent for individual policies and 15 percent for groups. If a company spent less than 80 to 85 percent of their premiums on patient care, they had to pay a fine. (Medicare has an overhead of 3 to 5.2 percent.)

As Elisabeth Rosenthal, author and editor of *Kaiser Health News*, learned, a higher medical loss ratio didn't dampen the amount the insurance company paid for patient services. Her book, *An American Sickness*, tells the story of a person whose infusions at an influential hospital (NYU) cost the insurance company five times as much as they would have had the patient received the same treatment at a nearby, presumably equally capable, facility. Rosenthal asked herself why the insurance company was willing to pay so much.

One possible explanation: Hospitals and physicians routinely overcharge, and insurance companies pay a negotiated portion of the bill. Then insurers show sick people the official bill and brag about the amount of money the person saved.

When insurers pay a higher portion of the bill and their profits fall below 15 to 20 percent, they are allowed to hike the price of their premiums. When people pay more for insurance, the company takes in more money and gets to keep 15 to 20 percent of the increased revenue.

The establishment of an "acceptable" medical loss ratio perversely rewarded insurers who drive up the cost of medical care and insurance. Health care premiums have risen 25 percent since Obamacare was enacted.

Two new taxes were enacted: a medical device tax and a "Cadillac" tax on high-priced policies.

The Trump administration established "association health plans." They still cover preexisting illnesses. They also "allow small businesses, including self-employed workers, groups that don't have many people with serious pre-existing problems, to band together by geography or industry and obtain coverage as if they were a single large employer." The plans "don't need the minimum benefits required by the ACA and they can drop maternity or mental health coverage." As the healthy are drawn away from the ACA, the exchange costs go up and the illness penalty returns.

At 1:30 AM on the morning on July 28, 2017, a bill that would have repealed Obamacare needed one final vote to pass. The person who cast it was Republican senator John McCain, a man who had recently learned he had a lethal brain tumor. He came to the Senate chamber, stood before his fellow legislators, and voted thumbs-down. No! The repeal of the law failed.

McCain was unhappy with the way Democrats had, years earlier, forced a "social and economic change as massive as Obamacare" through Congress. "Our health care insurance system is a mess. We all know it, those who support Obamacare and those who oppose it. Something has to be done." He didn't suggest repairing and saving the Affordable Care Act. But he spoke of "incremental progress, compromise, and chipping away at problems."

In 2018, Congress repealed the law's requirement that everyone must pay a penalty if they don't buy insurance. A federal judge in Texas decided that made the Affordable Care Act unconstitutional. He wrote that the ACA obliges people either to buy insurance or pay a fine. The requirement was a tax. When Congress repealed the individual mandate, the law stopped looking like a tax.

In August of 2018, the government started allowing states to take Medicaid coverage away from "people not engaging in work or work-related activities for a specified number of hours each month."

In December 2019, Congress passed a $1.4 trillion spending bill. It repealed the medical device tax ($20 billion less for health care in a decade) and the Cadillac tax ($193 billion saved by the wealthiest citizens in a decade). It also ended the health insurance tax that was used to pay for the federal and state marketplace exchanges ($164 billion over a decade).

The spending bill provided two years of Medicaid funding for Puerto Rico and other U.S. territories, and it barred HHS, the federal Health and Human Services department, from ending auto-reenrollment. They are not allowed to create a coverage gap for people who forget to enroll at the end of each year.

Atul Gawande once concluded that the United States may be the only developed country in the world where people are "unable to come to agreement" on the concept of health care as a right—on the idea that all should be able to benefit from the medical advances of the last hundred years.

As this book goes to press, the Supreme Court has yet to decide whether or not the Affordable Care Act is constitutional. When the case was argued before the nine justices, Chief Justice John Roberts said: "I think it's hard for you to argue that Congress intended the entire act to fall if the mandate were struck down, when the same Congress that lowered the penalty [for not buying insurance] to zero did not even try to repeal the rest of the act. I think, frankly, they wanted the court to do that. But that's not our job." Justice Brett Kavanaugh made a similar point. "It does seem fairly clear that the proper remedy would be to sever the mandate provision and leave the rest of the act in place—the provisions regarding pre-existing conditions and the rest."

In the spring of 2021, the court will decide whether or not the Affordable Care Act, Obamacare, is constitutional.

The health care we enjoy is only 120 years old. It gave a new heart to the tin woodsman, Dick Cheney, and helped the scarecrow, Joe Biden, keep his brain. (He had an aneurysm that bled.)

But we don't live in Oz, and there is no wizard. As detailed in the book our courageous lion, Lyndon provided socialized medicine to the elderly and impoverished, but when Hillary tried to provide health care for the rest of us, she was demonized. Bill Clinton was able to get HIV medicines for millions in Africa before the World Trade Organization changed the rules. Barack managed to squeeze the Affordable Care Act through Congress. Eight years later John McCain cast the deciding vote on a law that would have excluded many from operating rooms where the blind and lame are healed. He turned his thumb down and the act was defeated.

Americans cheered politicians when they decried socialized medicine, and cheered again when the same politicians told the hundred million insured by Medicare and Medicaid that they won't take their health care away.

We've allowed our insurance companies to teach our young that medical care is not a shared responsibility. Then many were surprised when the youth thought it unfair when we asked them to contribute to the cost of everyone's health insurance. We taught pharmaceutical manufacturers to not worry when they spend billions and buy other companies for their drugs. We will reimburse the acquisition cost and pay what the companies charge for the medications. We tell the poor of the world who can't remotely afford expensive medications and whose countries belong to the World Trade Organization—"suck it up." We tell the wealthy to buy stock and become part of the bonanza.

I wrote the book so people can understand where we are and how we got here. The future is yet in your power.

A NOTE ON SOURCES

A note on sources: In his bestselling book *The Great Crash of 1929*, economist John Galbraith noted "everyone needs to know on occasion the credentials of a fact." At the same time "there is a line between adequacy and pedantry."

I started medical school in 1958, was a medical practitioner for forty years, and started researching this book when I retired in 2010. During the last ten years I browsed hundreds of articles and magazines. Most of my sources are available on the web. A number can only be found in the books listed below or in subscription-only websites.

To avoid plagiarism I tried to put quotation marks when I used an author's exact words. I did not specifically footnote the quotations or the data, but my sources are all listed in the bibliography on my website: www.savingobamacare.com. (You can also find them at the original website, www.pondringdrugprices.com.)

Most web references are in URL form and that should make it easier for an interested reader to go to the website, copy the URL, and paste it in their computer's browser. Some of the sources are more difficult to check out because they come from subscription-only websites like: UpToDate, the *New England Journal of Medicine*, and the *New York Times*. Or they come from one of the books listed below.

SELECTED BIBLIOGRAPHY

I gained knowledge from the following books:

Alvord, Lori. *The Scalpel and the Silver Bear.* Bantam, 1999.

Angell, Marcia. *The Truth about the Drug Companies.* Random House, 2005.

Barnard, Christiaan & Curtis Bill Pepper. *One Life.* Macmillan Company, 1969.

Barton Smith, David. *The Power to Heal: Civil Rights, Medicare, and the Struggle to Transform America's Health Care System.* Vanderbilt University Press, 2016.

Blumberg, Baruch. *Hepatitis B.* Princeton University Press, 2002.

Breecher, Charles & Sheila Spezio. *Privatization & Public Hospitals.* 20th Century Fund Report, 1995.

Cohen, Jon. *Shots in the Dark.* Norton, 2011.

Covert, Norman. *Cutting Edge.* Self-published, 1997.

Doudna, Jennifer & Samuel Sternberg. *A Crack in Creation.* First Mariner Books, 2018.

Eban, Katherine. *Bottle of Lies.* HarperCollins, 2019.

Ferrara, Napoleone. *Angiogenesis.* Taylor and Francis, 2007.

Fredman, Steven. *Troubled Health Dollar.* Virtual Bookworm, 2012.

Fredman, Steven & Robert Burger. *Forbidden Cures.* Stein and Day, 1976.

Gawande, Atul. *Better.* Metropolitan Books, 2007.

Gawande, Atul. *Complications.* Metropolitan Books, 2002.

Hawthorne, Fran. *The Merck Juggernaut.* John Wiley, 2003.

Hughes, Sally Smith. *Genentech.* University of Chicago Press, 2011.

Kocher, Theodor. *Text-Book of Operative Surgery.* Adam and Charles Black, 1895.

Lindorff, Dave. *The Rise of the For-Profit Hospital Chains.* Bantam Books, 1992.

Loeck, Renilde. *Cold War Triangle.* Leuven University Press, 2017.

Miller, Wayne. *King of Hearts.* Random House, 2000.

Mueller, C. Barber. *Evarts A. Graham.* BC Decker, 2002.

Mukherjee, Siddhartha. *The Emperor of All Maladies.* Scribner, 2010.

Offit, Paul. *Vaccinated.* HarperCollins, 2007.

Pearl, Robert. *Mistreated.* Public Affairs, 2017.

Potter, Wendell. *Deadly Spin.* Bloomsbury Press, 2010.

Rosenberg, Steven. *The Transformed Cell.* Putnam, 1992.

Rosenthal, Elisabeth. *An American Sickness.* Penguin Books, 2017.

Silverman, Milton, and Lee Phillip. *Pills, Profits, and Politics.* University of California Press, 1974.

Starzl, Thomas. *The Puzzle People.* University of Pittsburgh Press, 1992.

Thomas, Lewis. *Lives of a Cell.* Bantam, 1974.

Vilcek, Jan. *Love and Science.* Seven Stories Press, 2016.

Wapner, Jessica. *The Philadelphia Chromosome.* The Experiment, 2013.

Ward, Thomas. *Black Physicians in the Jim Crow South.* University of Arkansas Press, 2003.

Werth, Barry. *The Antidote.* Simon and Schuster, 2014.

Werth, Barry. *The Billion-Dollar Molecule.* Simon and Schuster, 1994.

Additional major sources of information:

Fire in the Blood. A documentary movie about HIV, by Dylan Mohan Gray.

Hillary. A documentary (Hulu), by Nanette Burstein.

Obama's Deal. A documentary (Frontline) on the passage of Obamacare, by Michael Kirk.

CSPAN: Broadcast of U.S. House of Representatives committee hearing on prescription drug pricing, February. 26, 2019.

Correspondence with RoseMary De Moro, leader of the California Nurses Association.

CPSIA information can be obtained
at www.ICGtesting.com
Printed in the USA
FSHW010657160721
83152FS